TAKE THE

RED PILL

AND CURE
GLOBAL WARMING

TAKE THE

RED PILL

AND CURE
GLOBAL WARMING

A.G. CUSICK

Introduction

You know how it is; you go to school, go to Uni, get a girlfriend, get a job ...and then one day reality hits. This is the true story of one of those "reality hits" moments.

The seeds of this story were planted in 2001 by a series of articles in *New Scientist* magazine, entitled *Time to Rethink Everything*. A millennial State of the World review detailing the problems facing global humanity in the new century, and the possible solutions to these challenges.

It was a shocking read. When you take the time to sit down and study up on the immensity of the global issues facing humanity it is not only over-whelming, but terrifying. But I was as equally awed — by the potential of the solutions available.

Inspired, I spent time researching these topics ...before the demands of final year exams and then working life shifted my focus back to the immediate problems of everyday existence.

Soon, such abstracts as global poverty and global warming became as out of mind as they were out of sight.

...some years later changes in my life had led to my decision to travel around the world. And then one day reality hit. The abstract became the real and all of the "Big Questions" came to the forefront of mind once again. I knew then it really was *Time to Rethink Everything*; but the most pressing question to me of all: *Why?*

Why had all those potential solutions discussed years ago not been imple-mented? This was the question I found myself asking as arrived in Sydney, Australia after backpacking across Asia. As I returned to western culture, my culture, I realized that I was looking at the world in a new way.

To all those who helped make it happen

1

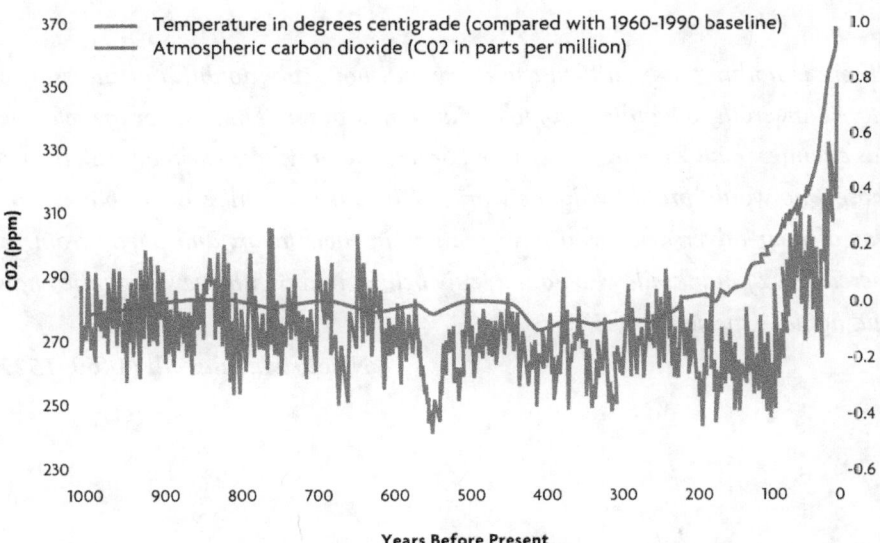

"There is nothing more difficult to carry out, nor more doubtful of success, nor more dangerous to handle, than to initiate a new order of things. For the reformer has enemies in all who profit by the old order, and only lukewarm defenders in all those who would profit by the new order. This lukewarmness arises partly from fear of their adversaries who have the law in their favor; and partly from the incredulity of mankind, who do not truly believe in anything new until they have had actual experience of it."

– Nicolo Machiavelli (1469–1527)

Contents

PART I:
The Journey External

Arrival

"We are in a fool's climate, accidentally kept cool by smoke, and before this century is over billions of us will die and the few breeding pairs of people that survive will be in the Arctic where the climate remains tolerable"

– Sir James Lovelock[1]

"And now I become Death, Destroyer of Worlds"

– Bhagavad Gita

Istepped out for a walk in the evening, away from the patio outside the hostel into the main street. Night certainly didn't flatter the place. The streets were dirty and full of litter, most of the fronts which weren't shuttered were sex shops or strip bars and prostitutes openly plied their trade on street corners. It was almost a parody, like some bad movie set. I had heard tale of Kings Cross before but couldn't believe what I was actually witnessing. Needless to say, the walk didn't have the bracing effect I'd been hoping for and I retired to my bed in the dorm room to lie down and think.

It felt like I was being crushed by pressure from every direction. The return to my own culture had brought up a homesickness, but I was almost as far from family and friends as it was possible to be. The inexpressible richness, exotica, novelty and wonder of travel in Asia had been replaced by the familiar in spirit but the distant in love. The sense of claustrophobia that came with either staring at the ceiling a foot or so above or out of the window to its panoramic view of a rain sodden brick wall ten feet away didn't help. Cold air streamed through the window's shutter, un-closable due to a broken mechanism, and rain drops sprayed off the window ledge into the room. It was a detail symptomatic of the shoddiness of the upkeep of the entire hostel, and I felt like I was holed up in some down and out in London, like so many over the ages, thoughts of glory pulled hideously to earth by the reality of trying to find work in the big city.

It had all started off so well. After an aerial journey which had taken me across the great red desert of the centre of the island continent, a luminous full moon dipped below the horizon and the aeroplane banked through coast-hugging clouds, revealing a spectacular view of the eastern seaboard and the city of Sydney itself. What adventure and excitement will this phase of my journey bring? I wondered. Even the welcome of grey skies, cold air and rain did not dent my mood of optimism. The girl at the desk of the hostel was called Kylie. Seemed like a nice way to arrive.

Kylie was blonde and slim and pretty and altogether Kylie-like but my mood began to fall as I surveyed the room in the hostel and contemplated the price that went along with it. After the in-expense of South East Asia, to get less for much more was not pleasant. The building was a characterless backpacker factory; functional but unkempt and untidy. There were piles of rubbish in the rooms and in the hallways. There were stains on the carpets, walls and ceilings. The reality of my situation began to sink in.

I had gotten some light relief from the burgeoning sense of pressure by talking to my roommates; Jan, an 18 year old Dutch lad fresh to travel and Keely, a pretty surfer from Saskatchewan of all places, who had recently been teaching English in China. The usual backpacker chat was easy enough. But I realised I was just going through the motions. I could not really relax. What's more, I wasn't really there. Questions had been flooding my mind since my arrival. What was I going to do while I was here? Where was I going to work? How would I meet people? Would I make friends? How much could I afford to travel? How long would my money last? What would happen if it ran out? What would happen if I didn't find any work? What would happen if I didn't make any friends? But all of these questions, pressing as they were, were just ripples beside the ocean swells of the deeper concerns that were preoccupying me.

No one was in the room while I was resting, and regardless no one could have seen me in my top bunk. But I covered my face. Tears welled up.

15/March

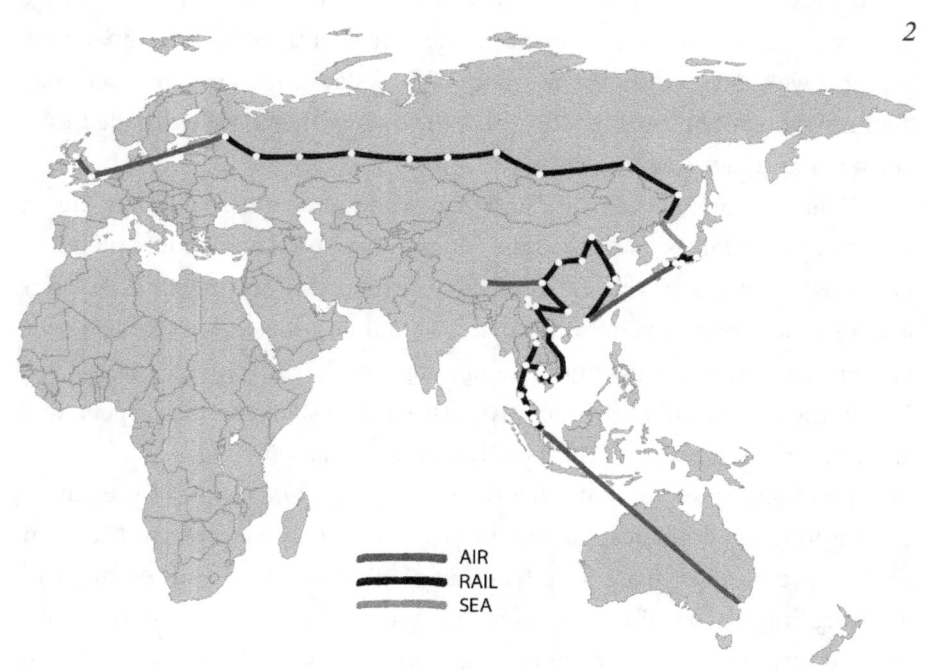

The Story So Far.

I spent the morning emailing friends. Contacting people I knew just brought home what I was feeling; homesick. Everything down from the grey skies to the brick buildings was reminding me of home. Even some of the streets had the same names. Yet I wasn't at home. I was about as far away as I could be. And I had to make decisions. I hadn't been working for almost a year now and it was make or break time for my CV. Get some contract work and polish it up while I was out here was the original plan. But the more I thought about going back to *tip-tap-tip-tap* every day the more I knew that I just couldn't do it.

Every single day for the past nine months had been like a new birth. I'd been experiencing life fresh and new with each sunrise. I couldn't stand the thought of spending six months in an office, pretending it all just hadn't happened. More than anything I wanted to keep moving, learning, growing.

The hostel patio was full of young backpackers sitting out, drinking and talking. The sight meant nothing to me. I didn't want to get involved. Drinking was out; regaining my health was my priority and that meant no habitual alcohol use. I had no interest in small talk either. There was just too much on my mind.

I walked into the city to try and get my head straight. I realised I was coming back down to earth. The familiar pressures, demands and routines. Expectations. I stopped by a bank to withdraw some money and the balance hit me like a punch in the stomach. There was no way I was getting round Australia on $300. I'd be lucky to last a month in Sydney on $300.

Even the sights and sounds of this new city couldn't take my mind off the yawning gap of what was to become of me. Even the simplest decision I made now would have ramifications that would stretch out throughout time, throughout the rest of my life. I realised I had reached a crossroads in my life. I walked up to the concert hall and looked out at the harbour bridge. It underwhelmed. It had not the significance of my reverie.

It wasn't just the little issues that were on my mind, it wasn't just the bigger issues of where my decisions would lead, there was something far more troubling, something deeper, and as I began to take in the culture around me, something omnipresent.

As I crossed the busy streets of the centre of town I perceived a monster; a behemoth astride and engorging the world. Each of those wheels that sped past me the articulations of the cogs in a planet sized machine. Observing the thunder and the smoke, I knew that every passing vehicle was mirrored by teeming others in every city in every country in every part of the world; each wheel around me representing hundreds of millions of other wheels, ever moving, never stopping. Business as usual. Compared to the marching machine of daily life, something so vast even my newly travel-stretched mind couldn't contain, so utterly unstoppable and relentless, I felt crushed, powerless and overwhelmed.

16/March

I resolved to get my new life under way. I got up early, showered, shaved and put on some smart casual clothes. An unexpected chink of light emerged during breakfast when my eyes alighted on a poster for the W.W.O.O.F. association on the wall beside the reception. Willing Workers on Organic Farms. I had heard of the movement before, when I had first planned my round the world experience; the WWOOF setup is a cultural exchange whereby a host, usually but not necessarily an organic farm, provides you with food and board, and in exchange you do four to six hours of work on their property. It is a working yet informal arrangement in the spirit of openness, exchange and international friendship. I'd never put any serious consideration into it until now ...but the poster had definitely got me thinking.

I performed the usual rounds; bought a mobile phone, set up a bank account, wrote up my CV (or Resume as it was known here), visited recruitment agencies and checked out the hostel notice boards. There was labouring work on the go, but I knew my delicate physique and introverted temperament wouldn't cut it on a building site. I would hold on that option until I had absolutely no other choice.

Setting things up kept me busy for the day, but my heart wasn't in it. As with talking to folk in the hostels, it felt like I was only going through the motions, putting on a front of normality and pretending to be interested in what whoever I happened to be talking to had to say. I was so withdrawn it was as if I was observing myself observe myself. Getting a job should have been a pressing need. But it all felt so tedious and inconsequential compared to what was really on my mind.

Should I just cut my losses and go home? One credit card bill was all it would take ...and then everything would be back to normal again. I could be home. I could be back to work in a few weeks. I could forget.

But the more I thought about it, the more such a course of action would be a slap in the face to all that I had experienced, all that I had learned, over these past incomparable months.

Just get through it.

State of the World

The developed world's debt is rising with no projected end in sight:

Debt (OECD)

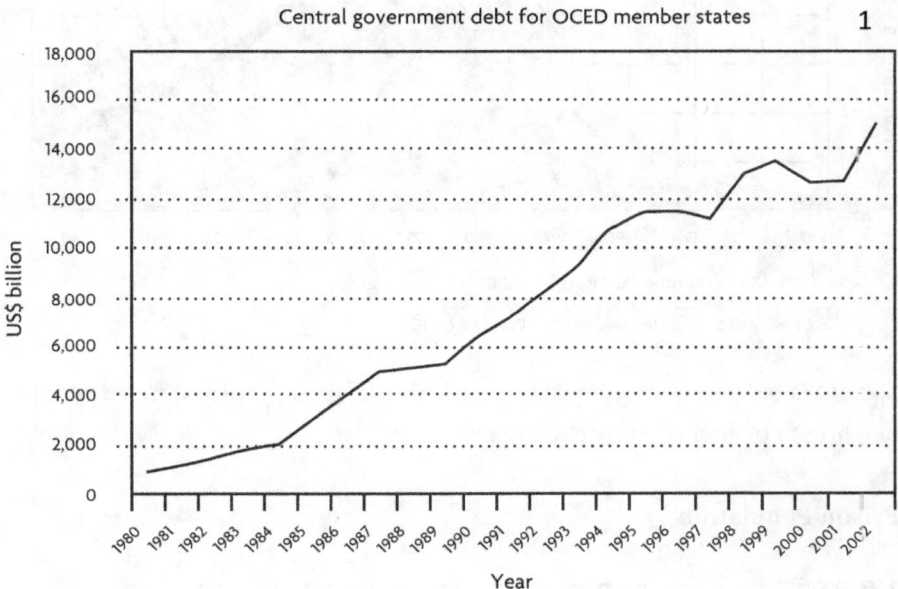

Central government debt for OCED member states

Meanwhile, the fossil fuels on which civilisation runs are expected to become increasingly difficult to find and extract, before running out:

Fossil Fuel depletion

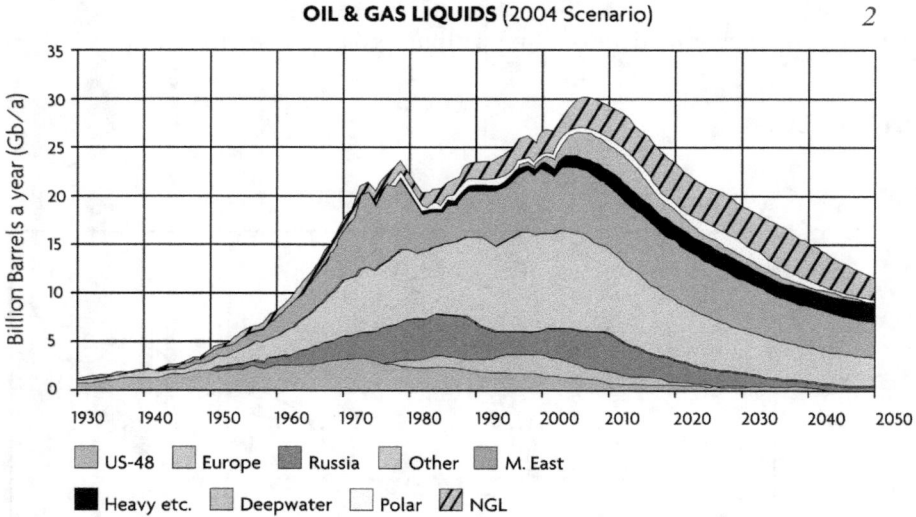

Society is becoming more unstable and violent, with more and more people needing to be locked away each year:

Prison Population

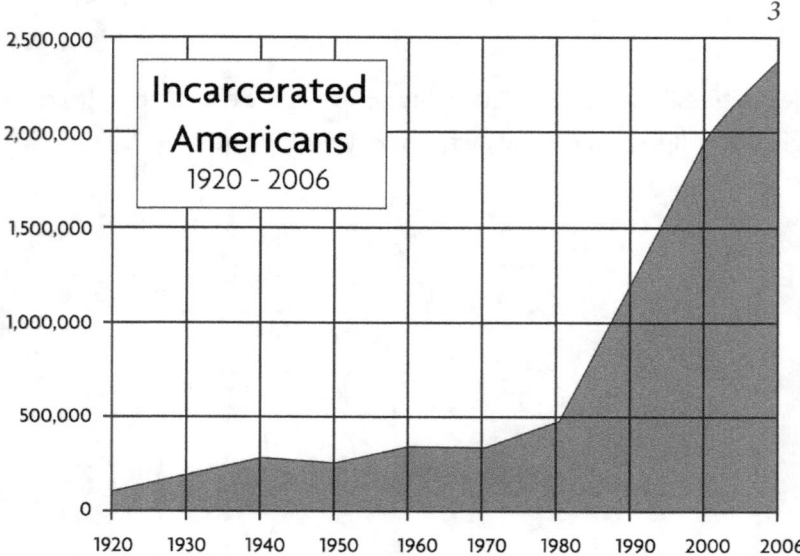

Fish stocks are collapsing around the globe:

Fisheries

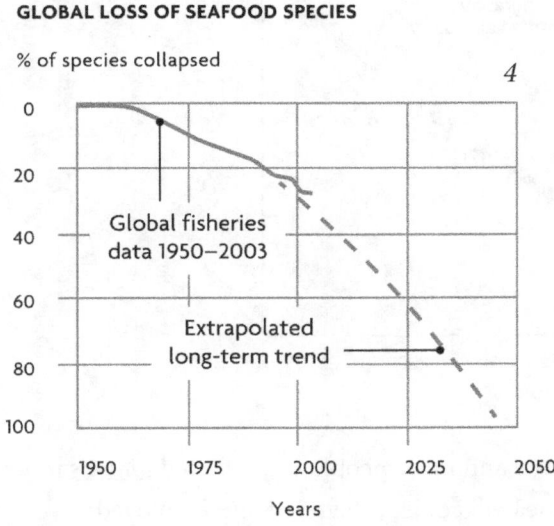

GLOBAL LOSS OF SEAFOOD SPECIES

% of species collapsed

4

Global fisheries data 1950–2003

Extrapolated long-term trend

Years

Many scientists now conclude that we are going through an extinction event like that which wiped out the dinosaurs, except that this time the culprit is us, and not an asteroid:

Extinctions

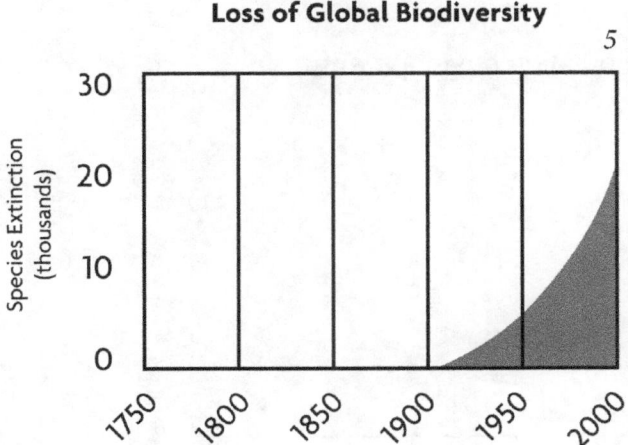

Loss of Global Biodiversity

5

Species Extinction (thousands)

The amount of Carbon Dioxide in the atmosphere is the highest it has been for 800,000 years, and contin- ues to rise year on year due to human activities:

CO$_2$

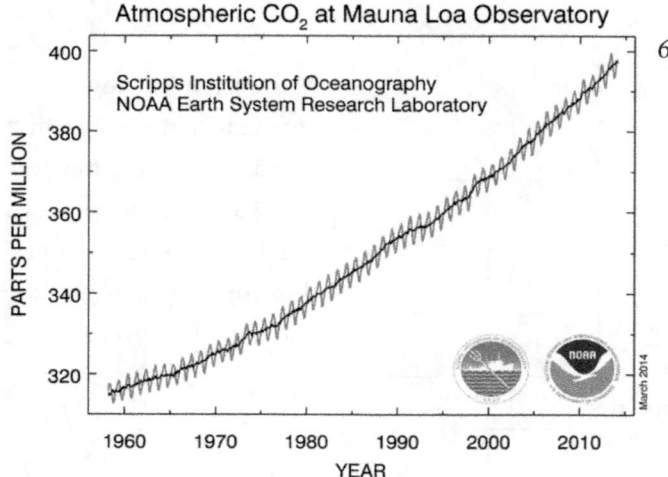

6

And into this picture comes more and more people — predicted *billions* more — to be fed, watered, clothed, heated, cooled, sheltered and employed:

Population

7

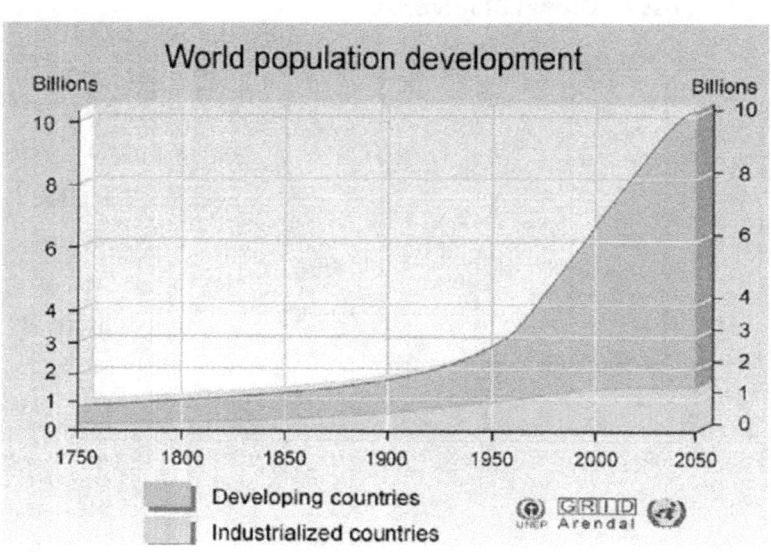

17/March

Sleep is nature's truest succour. I woke up on a wave of the most unexpected optimism and joy, seemingly from nowhere. I felt great and I didn't know why. Beyond the limin I must have had the best dream ever.

I bought the WWOOF book, which included one year membership and insurance for $48. I flicked through the list of hosts. Some properties in Tasmania caught my eye, as well as well as some closer to here in the countryside of New South Wales. Renewable energy, organic agriculture and eco-design featured in many of the properties. I smiled. It really was what I was looking for. I ticked off a list of interesting hosts around the country. One host not far from Sydney looked to have an ideal setup. It seemed it'd be a perfect way to introduce myself to WWOOFing and to take a break from my Kings Cross experience.

A new plan was beginning to form in my mind; I could combine my interest in sustainability with my vocation of travel. My heart leaped with the thought. One thing my travel experience had brought home was the idea of "sustainability". More than anything, I wanted to get hands on involved in the field after my travels. The WWOOF project could give me the perfect, practical learning experience and introduction I needed. It was also an opportunity to see places I would never otherwise have the chance to see, and meet people I would never otherwise have a chance to meet; off the beaten track in every sense. Things were looking brighter.

I was in a better mood. I joined up with Jan for shopping and our evening meal. An innocent abroad, only eighteen, his bravery impressed me. We discussed our plans here in Australia, and I became more and more enthused with the possibilities. Today had been a better day.

Goals:

Practice *qigong* everyday

~~Find contract work in old industry and work for several months, travel round Australia once free~~

Travel, work and WWOOF round Australia
> – Learn about renewable energy and all aspects of 'Sustainability'
> – Write book about my experiences
> – Solve Global Warming

18/March

Jan and Keely both checked out, and I was shifted rooms to another part of the hostel. This room wasn't an improvement though. The carpets looked like they had never seen a vacuum, the woodwork of the windows warped so that there was no option but to deal with the resulting draught. Some new room-mates from England arrived. I asked;

"So what do you think of this place so far?"

"Total Shithole!"

"Aye!"

But another day was another day to explore the city. I went for a wider sweep that usual, from the harbour bridge to Paddy's Market. The market was a lot less Irish than the name suggested. I'd always imagined Australia to be a pretty homogenous place, with a preponderance of blonde surfers, but turn certain street corners in Sydney and you'd think you were in Singapore.

My *qigong* practice was becoming a beautiful touchstone each day. I enjoyed walking down to the Botanic Gardens. It had a wonderful ambience and it wasn't far from the hostel. Occupying the curve of the bay east of the business district and the harbour bridge, it provided a beautiful backdrop to the Concert Hall. It was well maintained with a decidedly Victorian charm. Morning and early evening, this is where I went to practice. It was somewhere I could relax and enjoy the sunshine and the water view. Twenty minutes twice per day that I could switch off completely, lie in the grass and enjoy my *qi* flow. The only times I could forget about everything and just be.

A sense of unreality often strikes here Australia, which must have also been the experience of the early European settlers. So many similar yet different animals. So many unexpected creatures. Ibis padded around the streets as well as the pigeons, refugees from Alice's Wonderland. The crescent of the moon was tilted at a different angle. The constellations were unfamiliar. The seasons reversed. The swans black.

When you look up at the night sky and realise your friends and family are seeing something different; that's when you know you are far from home.

19/March

I woke from a disturbed sleep in which I dreamt of a Divine mannequin sitting in a room staring endlessly at a TV screen, oblivious as refuse piled up around it; slowly, gradually, relentlessly.

As I left the hostel for my morning round of *qigong* in the park followed by recruitment agencies I noticed that the Kings Cross casualties — the sick and the homeless, the drug addicts and the beggars — were out in force today. *Benefits*? *Methadone*? I didn't know, but it gave the place the look of a George Romero flick; shuffling shells of humanity occupying the streets in droves.

I occupied myself during the day with as much free stuff as possible; free internet ostensibly for job searching at the agency, free book browsing in the bookstores, free entertainment in the museums. All the while I read as much as I could, trying to get a handle on my unease. Endeavouring to turn my intuitions into facts. A sense of corruption had been building inside of me over the past few months. An intuitive sense that all was not well. It was beginning to bloom into my rational consciousness now; everything that I saw in the streets, everything that I read, sharpened my convictions.

On the way back to the hostel I was shocked from my reverie by:

"*You fucking wanker!*"

I was stunned, an assault from the blue.

I turned round to find an old homeless man, curled up on a park bench gesticulating furiously with one hand, the other clutching a paper wrapped bottle. I thought about walking on, but then decided I wasn't going to let such a violation of my person go unchallenged;

"Hey, what was that for?"

"You just fucking ignored me."

"I didn't see you."

Not that I would have given him anything if I had. I was about $200 away from destitution myself.

"Don't just fucking ignore me. I'm a human too you know...and...and... you're still a wanker."

He trailed off into further insults, and I turned around, disgusted. I was obviously to blame for all of the ills of his life.

After you've seen a man selling hot snacks on a street corner with a smile on his face, as I had done in Cambodia, a street corner he also sleeps on every night with his wife and child, it fairly changes your perspective on begging and homelessness in the first world.

I tried to shake the episode of, but my system was full of adrenal and I found it hard to let it go.

I want to get out of this place.

20/March

Typhoon Larry slams into Queensland.

It was already being anticipated as the costliest on record. Wherever I'd been on my global journey, freak weather events were breaking the trends.

I decided to use my new mobile to phone my first choice from the WWOOF book. The description in the book sounded completely ideal; a family of prize winning experts on sustainable building who lived on a property bordered by national parkland. It was near Sydney, but not too near. I was somewhat nervous about what would unfold. What if they rejected me or we didn't hit it off?

The host turned out to be an Englishman and we had a very cordial chat. They already had a WWOOFer staying but were free the following week and I was welcome to join them.

After the phone call I felt high; the plan was in motion.

21/March

I had used up my emergency supply of travellers cheques. Every withdrawal from my bank account was a haemorrhage. So it was with some relief that I found some removals work through the hostel notice board. I got picked up in a white van along with another backpacker, Dave, a big Geordie lad a few years older than me.

Dave was as sound as people from his city tend to be. Muzzy our employer, on the other hand, was a real Kings Cross character. An independent trader if ever there was one. Of mainly Turkish extraction, he cultivated the rakish guise of a man surviving on his wits against the odds. He was proud of his street nickname of "The Predator", a play on his dreadlocks; though I don't think it was intended as complementarily as he took it. He was a little crazy, but the sort of person who was alright if you were on their good side.

The actual work of removals was tough. Lifting cookers and freezers and tables up and down stairs all day was not what my relatively petite frame was made for. But when we'd finished I had made $165, which would keep me under hostel roof for a while longer. I was pleased, but Dave reckoned we'd been ripped off. He'd worked out that Muzzy, with his underground empire of white-vanning and porn vending, was likely making more money per week than the professionals he did removals for.

Dave recommended the place he was staying in over mine. It was a dollar or so cheaper per night so I took him up on his advice. I figured I couldn't do any worse.

22/March

Hostel II was pretty poor, but at least had a large free breakfast. Manna from heaven for the long term traveller. My room was a small, dingy three bed dorm. There was dampness running down one of the walls, but it still managed to be an improvement over my last one. The hostel atmosphere was better as well, so I decided to run with it.

One of my roommates was a decent English guy. Jason he was called, and he was in similar financial straits to myself. I chased my other roommate out of the window after it had made advances towards my person.

A very large *El Cucaracha*.

23/March

Kings Cross really was a sight to behold. A busy node on the transport network of the east side of the city, bustling commuters crossed paths with streetwalkers all hours of the day and all days of the week. The prostitutes here were different from those I'd met in Thailand. In Thailand the girls seemed to be able to withstand it more, perhaps because it was more of a choice through pressing financial concerns, but here the situation was more obviously the approaching terminus of a downward spiral.

A cute Asian girl approached me, her body language inviting;

"Mister, would you like girl?"

Her smile was supposed to be seductive but revealed her missing teeth, the consequences of heavy drug addiction.

"No thanks".

Her smile dissolved into a grimace and her face into a mask of furious anguish as she screamed abuse at me as I walked away.

So strange, I thought, for your entire being and it's emotions to be so focused on something outside yourself.

For a Yes/No answer from a stranger to flick a switch controlling your entire mind:

Happiness or Hate.

I had passed that girl often as her 'beat' was in the main road just outside the hostel. I saw the same show go on regularly with other passersby. Sometimes she would follow the ask/abuse routine that seemed common to all of the sick people in this place. At others she would dissolve following a snub into her own personal hell, her countenance contorting into an incarnation of rage and pain. She would start screaming. Screaming at the world in general but at no one in particular. On other days, no doubt having gotten her hit, she and her friends would be drifting hazily about in a deliriously good mood.

I contemplated how society could reach its Kings Cross nadir as I walked around Sydney. The great commodification of life increasingly devouring whole human beings. "Sex" confronted me wherever I looked. It poured from the cover of magazines and newspapers, was a suggestion occult throughout the world of advertising. A word seemingly disembodied and without context, floating ubiquitously on the very top of our consciousness.

Something you "Get". No longer something that you share.

Are you a human being or a light switch?

9

1	0
GOOD	EVIL
CHRISTIAN	MUSLIM
RIGHT	WRONG
RICH	POOR
FREEDOM FIGHTER	TERRORIST
WHITE	BLACK

25/March

I continued to spend as much time reading as I could. How can you just *get on with life* when the world is on course to be six degrees warmer? What does our day to day life actually mean with such a projected future? I was obsessed. I needed to get to the bottom of the issues of the day. I needed to find the solution.

The problem was that the more I read the further away solutions became. Every supposed solution became a problem, until I was mired in problems within problems, with no end in sight. And the more I learned, the darker my general mood became. Just as I thought I had found a book or an idea that promised to be an answer a deeper problem arose. There was also so much that was just plain contradictory or misleading. As I filled in the picture of the world, the gap between my expectations of society and the life that was going on all around me drew further and further apart.

This philosophical gap underscored the distance I was feeling from my fellows in the hostel. I wanted nothing more that to talk about these issues, but talk of deep issues requires deeps bonds and of these I currently had none. I was alone with my thoughts. I reflected ruefully that the closet thing I had to a friend was my roommate Jason, who worked nights while I was asleep and who slept when I was awake.

Corporate Globalisation

"The powers of financial capitalism had another far reaching aim, nothing less than to create a world system of financial control in private hands able to domi-nate the political system of each country and the economy of the world as a whole. This system was to be controlled in a feudalist fashion by the central banks of the world acting in concert, by secret agreements, arrived at in frequent private meet-ings and conferences."

– Carroll Quigley[1]

One of the most disturbing realities of our 21st Century world is the fact that an increasing number of decisions affecting our lives are being taken at supranational levels by international agencies and federal struc-tures which we, the electorate, have little access to or ability to influence. This decision making is in turn influenced both formally and informally by immensely powerful special interest groups, whose membership often pop-ulate the highest levels of western governments and international agencies.

It has only been in recent years with the arrival of the internet and the subse-quent global free flow of information that the background to corporate glo-balisation has come to light. Prior to this most of humanity remained largely in the dark, in no small part due to the fact that almost all of our mass media is either government controlled or owned by members of those very same special interest groups.

Unelected Governments

Bank for International Settlements (BIS)
Known as 'The Central Bank of Central Banks', the Bank for International Settlements was created under the official aegis of dealing with the repara-tions of WWI.[2] It's original ownership comprised host country Switzerland and the central banks of England, Germany, Belgium, Italy, France and Japan plus three U.S. private banks; J.P. Morgan & Co., First National Bank of New York and First National Bank of Chicago.[3] The stated purpose of the BIS is to

foster international monetary and financial cooperation and to serve as a bank for central banks.[4]

Despite its crucial position in world affairs, the likelihood is that its very existence is news to you — and this would be entirely in keeping with the history and the culture of the Bank for International Settlements; a most private of private organisations whose operations are conducted in secrecy.[5] A legally inviolable entity, it is beyond governmental jurisdiction and its membership is protected by greater than diplomatic-level immunity.[6] The apex of international banking, it is the main co-ordinator of global money supply along with the World Bank and the IMF.

Undemocratic Special Interests

The Bilderberg Group
The Bilderberg Group is an annual invitation only conference of elite members of society from Europe and North America. Named after the *Hotel de Bilderberg* in the Netherlands where the first conference was held in 1954, the meetings were incepted to foster understanding across the Atlantic divide. Members are sworn to maintain the secrecy — both of the agenda and of the participants — of the gatherings[7], indeed a certain Anthony C. L. Blair is one of many leading politicians who have failed to declare their participation.[8] Attendees include royalty, politicians, ministers, bankers, corporate leaders and the heads of media organisations.

Trilateral Commission
In 1973 an organisation along the lines of Bilderberg but including Japan was founded by David Rockefeller, chairman of Chase Manhattan Bank and former chair of the Council of Foreign Relations, and the academic Zbigniew Brzezinski.[9] Composed of a selection of bankers, intellectuals, politicians, corporate leaders and media heads representing the summit of their respective professions, the goal of the Trilateral Commission was to create a "New International Economic Order". The original Commission line up included a little known politician called Jimmy Carter, who rose meteorically to become President of The United States in 1976. Carter appointed no less than *eighteen*

other Trilaterals to positions in high office, [10] a preponderance that continues to the present day.

Contemporary membership of the Trilateral Commission includes representatives from corporations such as British Petroleum, Mitsubishi and Chevron, banks including Citigroup, JP Morgan Chase and Goldman Sachs & Co., and media conglomerates such as Atlantic Media Company, Newsweek International and AOL Time Warner.[11]

Unaccountable International Agencies

The vehicle for the "New International Economic Order" has been the U.S. executive, the Federal Reserve, the BIS and the international financial agencies descended from the institutions created at the Bretton Woods conference of 1944:

- **International Monetary Fund (IMF)**
- **World Bank**
- **World Trade Organisation (WTO)**

The ideology behind these agencies, termed the *Washington Consensus*, and the process by which they restructure the economies of indebted nations, *neoliberal economic reform*, have been the subject of widespread international concern and condemnation.

An Insider Critique

Founded in 1944 to rebuild post-WWII Europe, the World Bank and IMF today find their major involvement in Third World development and debt repayments, respectively. To highly indebted nations they are the lenders of last resort, but their loans and debt assistance come with strings attached; 'Conditionalities' and 'Structural Adjustment Programmes'. Investigative journalist Greg Palast interviews former World Bank chief economist Joseph Stiglitz in his book *The Best Democracy Money Can Buy*. Together, they dissect the mandatory four step assistance program[12] the World Bank and IMF apply to all developing nations receiving loans:

1. Privatisation

The required privatisation of state industries;[13] changes which produce an immediate rise in unemployment and are often accompanied by large cuts in remaining public sector jobs and wages. Former national assets — ranging from national banks, to oil and mineral resources, to public infrastructure such as telecommunications, water and electricity — are then open to being asset-stripped by multinational corporations and international banking conglomerates.[14]

2. Capital Market Liberalisation

The lowering of barriers to the movement of capital in and out of the nation in question. Theoretically, this creates the conditions for international investment, in practice, it often quickens the flow of capital out of an already weak economy; a process called the "Hot Money" cycle — at first, speculative investment arrives in markets such as currency and real estate but then exits rapidly if conditions deteriorate. To tempt international speculators back to the affected nation, the IMF demands massive interest rate increases which have the effect of crashing property prices, crippling domestic industry and further depleting national savings.[15]

3. Market-Based Pricing

Currency devaluation and elimination of exchange controls feature in Structural Adjustment.[16] This has the effect of "Dollarizing" domestic prices[17] and amounts to raising the cost of essentials such as food, water, fuel, electricity and medicine.[18] Understandably, such changes often result in enraged public protesting. Chillingly, these protests are expected outcomes of the assistance plans, where they are euphemistically predicted as "social unrest". There have been multiple examples of these 'IMF riots' such as the case of Ecuador, which erupted following an IMF directed 80% increase in the price of cooking gas. The suppressing forces of police and tanks which responded to the situation were also prescheduled by the IMF as "resolve".[19] Rioting and associated disruption further escalates the drain of investment out of a nation and often leaves the few remaining assets in a country to be purchased by multinationals at minimal prices.[20]

4. Free Trade

"Free trade" requires the removal of all obstacles to trade across the nations' borders. Of course, the nations of the First World do not follow their own "free trade" prescription, and maintain their own protectionist policies, including extensive subsidies for agriculture and industry, tariffs on imported goods and legally protected monopolies under the rules of the WTO. Stiglitz and Palast compare this enforced one-sided opening of markets to the Opium Wars of the 19th century.[21]

The combined effects of the neoliberal shock doctrine are obviously catastrophic; wages crumble and unemployment surges as the price of daily essentials goes through the roof. These changes in turn backlash on the domestic economy in a vicious cycle. Social investments including primary healthcare and basic education are cut to the bone and the savings redirected to the servicing of external debts. It is additionally distressing to remember that these programmes are enforced by the lenders of last resort on nations *which are already impoverished*.

These measures have been implemented in more than 150 indebted countries around the world and have been instrumental in bringing about the conditions that have led to numerous modern famines and civil wars, including the genocidal collapse of Rwanda and the outright destruction and fragmentation of Yugoslavia[22]. Perhaps due to the amount of criticism they have drawn, 'Conditionalities' have been renamed 'Poverty Reduction Strategies'.

- **Just how effective has the "New International Economic Order" been in reducing poverty?**

If we look at the incomes of the worlds elites, we can only say that it has been an unparalleled success; they have been having their poverty reduced considerably since the 1970's, to the extent that at the end of the last millennium the worth of the worlds 358 billionaires exceeded the combined annual incomes of countries accounting for 2.3 billion people, or nearly half (45%) of the world's population. In the years from 1960 to 1991, the richest 20% increased their account from 70 to 85% of global income, while the share of the poorest 20% decreased from 2.3 to 1.4%.[23]

It was in the face of such devastating statistics that the head of the United Nations Development Panel was moved to warn, "if present trends continue, economic disparities between industrial and developing nations will move from inequitable to inhuman."[24]

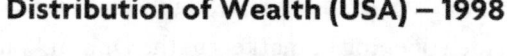

Distribution of Wealth (USA) – 1998

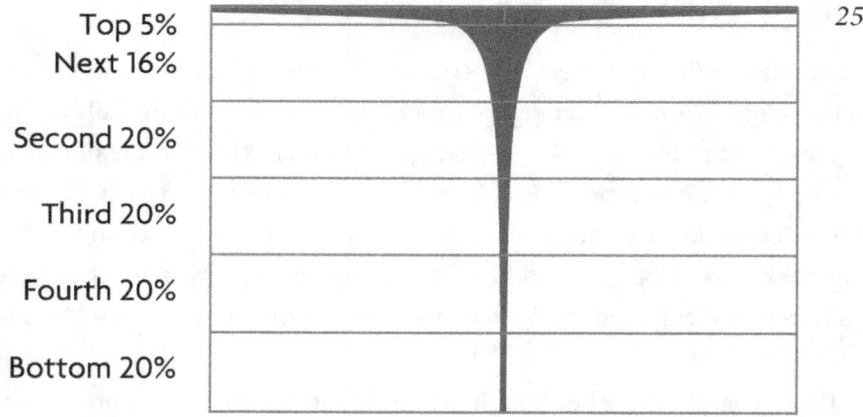

25

Top 5%
Next 16%

Second 20%

Third 20%

Fourth 20%

Bottom 20%

As between nations, so within nations; Globalisation has contributed to a similar polarisation of wealth in the First world. At the turn of the millennium, the top 10% of households in the United States controlled 75% of the country's wealth, with the bottom 90% holding just 25%.[26] To put this further into perspective; the top 1% owned almost 40% of the country's wealth, with the bottom 40% of the entire citizenry holding just 0.2%[27]

- **How successful have the lenders of last resort been in stabilising the economies of the third world?**

In 1970 the outstanding debt of the world's 60 poorest countries (LDCs) stood at $25 billion, by 2002 this had increased to a staggering $523 billion. Over the course of this time period those nations had reimbursed their creditors to the sum of $550 billion of principal and interest.[28]

While the majority of the population of planet earth sink deeper into this perpetual debt servitude, the estimated yearly turnover of the world trade in illicit drugs compares favourably with the LDCs outstanding debts at $321 billion in 2004,[29] an annual resource flow dwarfed by global military expenditure which stood in the same year at $1.1 *trillion.*[30]

Despite the obvious asymmetries they promote, our international agencies and their neoliberal dogma have many silver-tongued promoters within the mainstream media, but perhaps the best summary of their nature comes from the horses' mouth so to speak; to quote geostrategist, former U.S. National Security Advisor and Trilateral co-founder Zbigniew Brzezinski:

'In addition, one must consider as part of the American system the global web of specialized organisations, especially the "international" financial institutions. The International Monetary Fund (IMF) and the World Bank can be said to represent "global" interests, and their constituency can be construed as the world. In reality, however, they are heavily American dominated and their origins are traceable to American initiative, particularly the Bretton Woods Conference of 1944.'[31]

Further Reading

Mortgaging the Earth by Bruce Rich
The Globalisation of Poverty and the New World Order by Michel Chussudovsky

26/March

As I left the hostel after breakfast I had to skirt a police cordon in the main street. Turns out that someone had been gunned down with a shotgun in the toilets of a nightclub.

Just another day in Kings Cross.

John Perkins', *"Confessions of an Economic Hitman"* was one book I was reading which really resonated. After my prolonged period of travel I had seen some of the sights of inequality in the world that had so affected the author, conventional global economics as a source of destruction was something I was beginning to wake up to and the spiritual nature of the message of the book, perhaps unbelievable to some, certainly made sense in light of my recent experiences.

Returning to Kings Cross every evening always brought the abstract home. There was a sign in the main street above a sex shop which read:

"Love Palace"

Love.

Love could never be further from this place.

27/March

Away from the Cross, I enjoyed walking around Sydney. It kept me going. The Botanic Gardens were beautiful, full of flowers and plants exotic and native. The Opera House and the harbour view spectacular. Most days the weather was good.

But even practicing on green grass underneath majestic banyan trees with a view of the grand natural harbour, I still couldn't escape. It was always there, tugging at the back of my mind. The name of my place of practice gave it away; *The Domain*. Domination. No longer belonging to the original inhabitants through force of arms. The drivers that had shaped this culture, my culture, out in the open and unrecognised. Hidden in plain sight.

28/March

I was glad to be getting out of Sydney and my last practice a wonderful way to leave.

My flowing meditation was deep and peaceful. I was extremely relaxed. I could feel a tingling sensation all over. Then a pulse of energy as a massive bubble of *qi*, like electrical plasma, pushed through a blockage in my right thigh. It was a part of my body I'd been troubled by sciatic numbness and pain in for years. I was awed as the air around me resolved into stars.

Escape from Kings Cross

29/March

There was another WWOOFer due to join up with the family, and so we arranged to travel up to our host's remote locale together. I met up with Tony at a train station several miles outside Sydney. It was my first encounter with a fellow WWOOFer and I couldn't shake the mental preconception of someone equipped with natty dreads. Tony turned out to be a dread-free and completely normal looking Australian bloke, a Ph.D. student who was planning on interviewing our hosts as part of compiling his thesis.

It turns out he was having a bad day; he had left his car keys in Sydney and was currently borrowing his friend's spare car. Just as he was explaining this to me the engine spluttered and died. His bad day continued. We pushed the car off the highway and called a tow truck. Tony's friend Keith — the owner of the borrowed car — eventually came to pick us up. Such is the incestuous nature of hostel life, listening to the two of them was my first experience of proper Aussie banter. There was a bit of stick but all in the best of humour. Keith seemed to be taking events in his stride and his first suggestion when we got into his other car was:

"Bottle shop!"

That most Australian of inventions: a drive through off-licence! The unexpected development meant that we would have to forgo travelling to our WWOOF hosts today and instead crash overnight at Keith's girlfriend's house in The Entrance, a suburb on the coast, where Tony's key-less car happened to be parked.

Keith and his girlfriend Isabelle made us feel welcome, and we went for an evening's swim at their local outdoor seawater swimming pool before a fantastic dinner and a night in watching *The Trailer Park Boys*. It was my first taste of what 'real life' was like out here in Australia, and I could certainly see the attraction. It was a pleasant evening and they were lovely people, and I reflected on what an unanticipated boon the breakdown had turned out to be.

30/March

In true WWOOFing spirit we helped around Isabelle's house and garden in the morning. Tony's car keys arrived by post in the early afternoon and we recommenced our journey once more.

We drove towards our destination through progressively more rural and then more sylvan scenery. My first real sight of the Australian landscape, it was a strange hybrid of the alien and the familiar to my European eyes. The land rolled in to the foothills of the Great Dividing Range, where dairy pasture lapped against the remnants of primeval Australasian forest. Tony and I were both struck by how lovely it looked, as we seemingly travelled back in time. I sighted my first kanga-thing, bounding away from our oncoming vehicle, although it turned out to be a Wallaroo rather than the larger more famous variety. The roadway became un-surfaced as we moved deeper into the hills, and we eventually made it to the drive that would take us up to the families' property.

William and Samantha's property was impressive, an example of a modern environmentally friendly building, but reminiscent of a large old country farm house. Samantha greeted us at the door and I was introduced to Tim (14), Ivy (13) and their very enthusiastic dog, Cocoa. We met up with William inside over tea. William's icebreaker:

"What's your passion?"

"Sustainability" I answered, feeling glad to have found other people who shared my enthusiasm.

Such is life on a semi-self supporting property that shortly after our welcoming meal we were straight into the heavy stuff. We discussed the possible WWOOF projects for the week and settled on the construction of a stable for Ivy's horse. William, Tony, myself and the kids drove up in their jeep to the copse they used to source firewood and building materials. Two gum trees of the required dimensions came down and we set to work cutting firewood and stripping bark. Despite my recent heavy lifting and epic walking I found the manual labour tough but enjoyable work. I noticed that William and his son Tim, despite one looking like a university professor and the other being a 14 year old boy, had a degree of strength and fitness that belied their exteriors thanks to their outdoor lifestyle.

The rest of our day was taken up with wood manoeuvring and cutting. William, Samantha and Tony had bonded well during the course of the day and were sharing deep life stories by tea time. I sat in as Tony taped an interview with them for his project files.

William and Samantha ran mud-brick workshops, with William being a qualified architect with over twenty years experience in researching and designing sustainable structures. They had designed and built their own house, which was heated and cooled by the passive solar design of their house — simply by the nature of the mud-brick building material and the ventilation of the windows. Their toilet system was self-composting and their solar hot water system an ingenious no moving parts design involving a raised water tank and a metal heat sink. The only part of their setup that they hadn't self-designed and self-built was the electrical supply in the form of an array of sun-tracking solar panels stationed outside the house.

As we sat round the table for dinner, I became conscious of the fact that I was now officially under the roof of "alternative" people — people not like those I had come from. I worried about what *faux pas* I could accidentally make. But as dinner progressed, religion, politics, vegetarian diet and mud-brick housing choice aside, it became clear that they were a family like any other — warm, welcoming and down to earth.

Bedtime arrived and I got first choice of place to sleep as I was staying longer than Tony. It was a choice between a little mud-brick chalet or a caravan. My childhood impressions of Australia came to mind:

"What's got more spiders?"

"The Chalet"

Rather ungallantly:

"I'll go for the caravan then"

After saying goodnight I stepped out into the pitch darkness so familiar to the countryside yet unremembered by the city dweller. The sunny, friendly valley with the wooded hills all around might have become sinister and haunting had I actually been able to see anything at all. Fear in the human mind is never far away. Eventually my eyes adjusted enough to make it to the caravan, where I hunkered down under several layers of bedding for a fitful night's sleep.

31/March

Mornings in March in the valleys of the Dividing Range can be chilly I discovered. After ten months of constant summer my body didn't enjoy the readjustment to cold one little bit. It was a very early start as well, but I wanted to make an effort to fit in with the routines of the family.

A large breakfast of porridge and fruit with tea started the day. After some morning wood cutting Tony interviewed me for his project. He left soon after, and it was hugs all round as he said goodbye. Myself, Tim and Ivy continued on with the business of digging post holes for the stable.

Mud-brick House

- Passive solar (stable inside temperature; hot in winter, cool in summer)
- Self construction (self-design + option to self-build)
- Simple to build (mold construction + in-situ earth)
- Fits aesthetically with local surroundings (in-situ origin)
- No transport or material costs (only CO_2 comes from food to power builders)
- Durable and long term
- Yet simply and safely deconstructed when no longer needed (no pollution)

Solar Hot Water
- Self built
- Self-circulating, no moving parts

Composting Toilet
- Self designed and self built

Solar Electricity
- Sun-tracking Solar Panels
- Grant from government towards cost
- 1.5 kw at peak

In the afternoon I got some time off and a chance to look around their prop-erty — the house was set in a flat grassy valley that was itself a spur of a long valley which the access road ran along. The long valley's width varied along the length of the property and contained a low creek bed, which was currently dry. They owned over three hundred acres in all, running up to the top of the surrounding wooded hills and about one mile down the road in each direc-tion. It was a remote and beautiful place in the midst of the Great Dividing Range.

William was English by birth and Samantha a Kiwi and they had moved to Australia as they had found it the best place to purchase land for the twin purposes of habitat protection and the development of sustainable community. Their creations and achievements were both stunning and original. The mud-brick house itself was wonderful. That very term "mud-brick" doesn't make for a pretty mental picture, but the building material gave the place the look of a tastefully finished roughcast conventional house on the outside and inside, with glazed mud-brick floors, it was reminiscent of a traditional stone cottage. Airy and open plan within, there was a large kitchen and living area, a dining area and a stairway to the rooms upstairs. A vintage wood powered stove pro-vided energy for cooking and auxiliary heating.

Amazed by their setup, I immediately thought of the potentials for the developing parts of the world. Samantha told me that the whole family had lived in Africa on a charity mission to teach their methods of self-building mud-brick housing and solar hot water. However, I was surprised to find that they had had a mixed experience as part of the development aid industry, which they summed up as:

"People living lifestyle A telling people living lifestyle B to live lifestyle C."

01/April

I felt very at home with William and Samantha and their kids. Samantha had commented that a lot of WWOOFers can get homesick in Australia and enjoy being back in a family situation. Though sometimes the opposite happens and the WWOOF situation triggers a homesickness! The cultural similarity of Australia had definitely brought up a homesickness in me, and I was indeed enjoying a return to a family atmosphere. They were just like any other family with children approaching their mid-teens. Tim and Ivy were happy kids but got on like chalk and cheese, one with the quiescence of the newly adolescent male and a love of teasing his sister, the other talkative and curious and fond of horses and not fond of being teased by her brother.

They had chosen to home school their children and I could see from the WWOOF book that a lot of the hosts had made the same decision. Most of them had made their choice for philosophic reasons but practicality also played a part in the huge expanses of Australia where the distances involved in public schooling could be impractical or expensive.

The horse pen was taking shape, and when I wasn't involved with that there were a range of other tasks on the property to keep me busy such as cutting firewood or picking fruit. Compared to Kings Cross, it was heaven. My routine was roughly about four hours outdoor work per day and a little while cleaning up pots and plates after each meal. I had plenty of time to borrow books, walk about and explore their land, play with the irrepressible Cocoa and enjoy my *qigong* practice in ideal natural surrounds. It took time to heat up in the morning but by the afternoon when I had time off it was always warm and sunny. Hemmed in by two national parks it was a beautiful setting under bright blue skies, and I felt revitalised after the shock of landing in Australia and trying to work out what to do with my life.

In the evening we drove into the bright lights of Wollombi. Once a famous stop on the settlers' overland trail, it was now the very definition of sleepy backwater. There was a community play on at a small local theatre and we got free entrance as Samantha was a member of the catering crew. I wasn't expecting much but we were treated to a truly brilliant amateur production of Oscar Wilde's '*The Importance of Being Earnest*'. On the way back home a wombat stepped out of the bushes onto the road in front of the car and then froze, dazzled, by the headlights. We stopped. It stayed there. We moved forward a bit. It stayed there. We moved forward a bit more. It got part of the message, but ran backwards along the road *in front of the route of the car* rather than off to the side and safety. After a few minutes of wombat pursuit the road curved around a bend and the little fella, still running straight forward as fast as his short legs could carry him, plunged over the edge into the creek bed in a rustling crash of bushes. Not the brightest, marsupials.

02/April

The stable had been finished and Ivy's horse moved up from its paddock on a neighbouring property. She was as happy as a thirteen year old girl could be.

Like a lot of the hosts in the WWOOF book, their property wasn't a farm, and they didn't grow all of their own food. They had found that the bush here was just too full of life, and that growing a lot of produce would mean taking measures that they didn't want to take against the wildlife. The vegetable patches they did have needed a lot of protection against the possums and wallabies. But that didn't mean there was a lack of things to do; 300 acres of land needs to be managed. I got on with mulching round the trees in the garden and mowing the extensive lawn space.

In the afternoon, Samantha and the kids set off for town for a few days. As part of the homeschooling system here they had to attend occasional refresher courses in the main town, which I assumed was to keep track of their progress and provide them with materials. They were planning to make the most of the excursion and were visiting friends while they were away as well.

03/April

With just me and William about now, I had a lot of opportunities for one-on-one discussions. Our exchanges were always enlightening; that ancient archetype of teacher and student. William had been thinking about sustainable buildings and lifestyles since the 70's and was hugely experienced. Brimming with enthusiasm as I was, I realised I was a complete beginner in comparison, only now waking up to problems glaringly evident 30 years ago. His philosophies of building, although completely leftfield at first glance were both beautifully thought out and practical.

One thing he taught me was that the energy costs involved in modern construction materials, be it concrete or steel, are huge. Until that point I had tended to think in terms of transport for energy use and carbon dioxide emission, but in modern industrial society there is simply no end to the everyday things that have a hidden energy and therefore greenhouse gas cost.

The family deliberately didn't have TV but they did have a great DVD collection which I could enjoy in the evenings in the main house. The extended editions of *Lord of the Rings* were an atmospheric choice in this remote sylvan place.

04/April

Today was the WWOOF task of clearing alien plant species. As we surveyed their property it became obvious that this was a trial that would challenge Hercules. The species had become endemic, and all the clearing we could manage in several hours was only a token effort towards reduction rather than a step towards elimination. Yet another lesson.

I explained to William as we worked that the combination of my travel experience and the literature I had read during it had completely opened up my mind; a complete conceptual shift of my perspective on humanity and the environment, leaving me with so many unanswered questions. At each phase on my journey of contemplating "sustainability" I thought I had a mental grasp of the answers, only to sink deeper and realise problems behind my apparent answers. He agreed that once you *got it*, the whole picture was interconnected and holistic:

"It is all about **base assumptions**"

It was quite common in his experience to try to introduce others to larger perspectives only for them to mentally go down one particular highlighted route wearing the same blinkers as they had previously. I tried to get to the bottom of the issue:

"At the end of the day...I think it comes down to personal responsibility."

"Personal growth" he corrected.

05/April

Time to go. William dropped me off in town for my bus journey back to Sydney. I thanked him for the wonderful experience of staying on their property, and asked him to pass on my regards to the rest of the family.

Return to Sydney

Spending time with William and Samantha's family had left me feeling revitalised. I congratulated myself on the decision to WWOOF. The experience had been better than I could ever have imagined; meeting a wonderful family, having some great fresh food, experiencing the richness of the wilderness, learning so much about sustainability. My goals were beginning to take shape. I felt renewed and excited; now that I'd seen what the real Australia had to offer I couldn't wait for the rest of the journey. I was even looking forward to going back to Sydney — I needed money to fulfil my plans and Sydney was the best place to get it.

06/April

The nurture of the known had brought me back to the devil I knew, King's Cross, and my last hostel. A familiar face when you're alone on the other side of the world makes all the difference. I slipped straight back into my routines of week's past — up early, free breakfast, train in park, scour agencies, study, train in park, hostel, dinner, hang out.

10/April

I had a daily competitor for who could eat the most at the 6.00 a.m. breakfast shift, a Chinese gentleman who also wasn't in the mood for talking at that time in the morning. I was painfully aware of the distance between myself and the others in the hostel. I tried to work out what it was and kept coming back to that *philosophic gap*. I sat eating my breakfast watching the morning financial news, mind boggling at the global mass credulity involved. It felt like someone was injecting poison into my eyes and ears. *No wonder the world is so fucked up*, I thought, *when we live constantly in a sea of lies*.

12/April

During my recruitment agency rounds I saw an ad that made my heart leap:

"Get fit and earn money"

Temporary work right up my alley. I wanted to get fit, and I wanted to earn money. I applied straight away.

13/April

I had been reviewing the costs of transport around Australia.

They were huge. And when you looked at the distances involved it made sense — you weren't trying to get round a country, you were trying to get round a small continent.

With my travel plans sketched, I knew I had to make a big decision on my apartment back home.

The more I reflected on it the more to sell felt like the right decision. I would need financing for my book and the change of career direction I wanted to make when I got back. I would need time to re-adjust, and to write, which I would get by moving back in with my folks.

14/April

Conversations with Jason, in those few overlapping hours when we were both taking part in waking awareness, were my only release from the detachment I was feeling. He empathised with my financial circumstances, and had been in the same situation before finding work in the factory.

16/April

I attended church for Easter Sunday. Why? I felt the reality of Spirit more than ever and I guess I wanted to participate in community expression. Instead, that building corruption I had been feeling was brought into stark relief.

What was this all about?

The teachings of a great spiritual leader partially lost in time, welded forc-ibly to an Imperial Institution and nailed to some extraneous fairy tales. As I looked around at the building and the people within reciting by rote I no longer felt the same comforting sense of the greater me that I once had. The whole structure was a statement of empire, even down to the arrangement of the pews; regimented and militaristic.

What would Jesus have made of all this? I thought.

I found it remarkable that differing interpretations of the life of Jesus had been almost completely expunged from the mind of the average church-goer.

What other paths of development could this faith have taken?

Could our culture have taken?

I wondered.

Sick Money

"Let me issue and control a nation's currency and I care not who writes the law"
– Mayer Amschel Rothschild (1790)[1]

Sometimes the most difficult things to find are the ones which turn out to have been right in front of you all along. For me, money is the ultimate case in point — earn it, save it, spend it ...but do we ever question it? Our life often revolves around this most invisible of material base assumptions.

An excellent exposition of the history of our monetary system, its consequences, and the revolutionary potential of solutions can be found in Deirdre Kent's *Health Money, Healthy Planet.*

History of Banking

Money has come a long way from the barter and gift exchange in prehistoric societies which birthed it. The most significant change occurred during the middle ages when European goldsmiths began printing paper receipts for loan amounts of gold. Eventually these receipts began to circulate among communities as token agreements in their own right; essentially an **IOU**. Observing this process taking place, perspicacious goldsmiths calculated that the likelihood of all debtors collecting at the same time was very small — so by printing more receipts than they had gold in their vaults they could thereby increase their profits manifold.

Their next discovery was that through control of interest rates they could 'row' the economy; at low interest rates more people took out loans and money supply expanded, but as interest rates were raised money supply correspondingly contracted and an increasing number of loans could not be repaid — thus the goldsmiths profits were further enhanced through the seizure of defaulted assets.

This practice was legalised and monopolised with the creation of the Bank of England in 1694 and thus what began as no more than a confidence trick now provides the foundation of our modern banking system; **fractional reserve banking**.

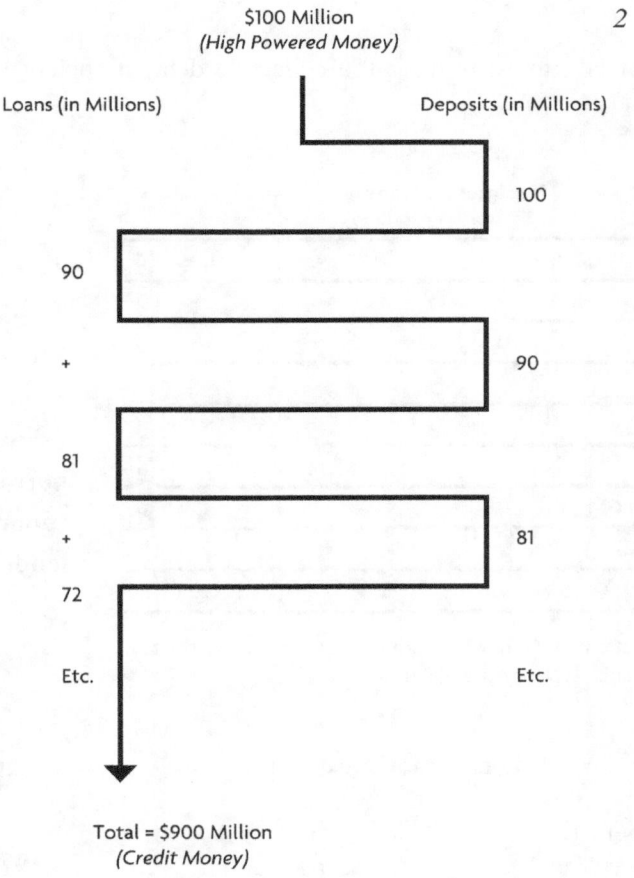

Total = $900 Million
(Credit Money)

Most of us take for granted that money is created as interest bearing debt, for example when we take out a mortgage loan to pay for a house, but we may not realise that the loans we withdraw are *not backed by deposits* within the bank but instead the money is **created out of nothing** by the banks themselves. The catch is that the *interest we must repay is not created* and so there will always be a shortage of money to repay debts within society. In order for a society to completely pay back its loans some of its members must take out further loans, thus creating a debt spiral.

Consequences of our system:

1. Money and debt are created simultaneously.

As this debt is interest-bearing the collective debt of society must always increase.

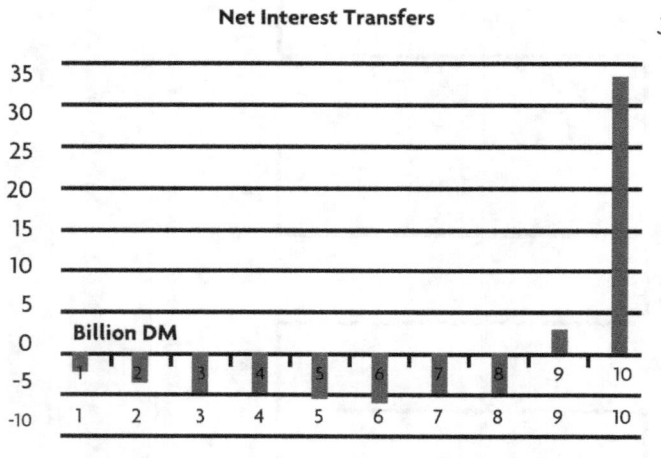

Net Interest Transfers   *3*

2. **Perpetual resource transfer from net borrowers (poor) to net lenders (rich).**

Systematic wealth transfer from the bottom 80% to the top 20% of society, Germany, 1982.

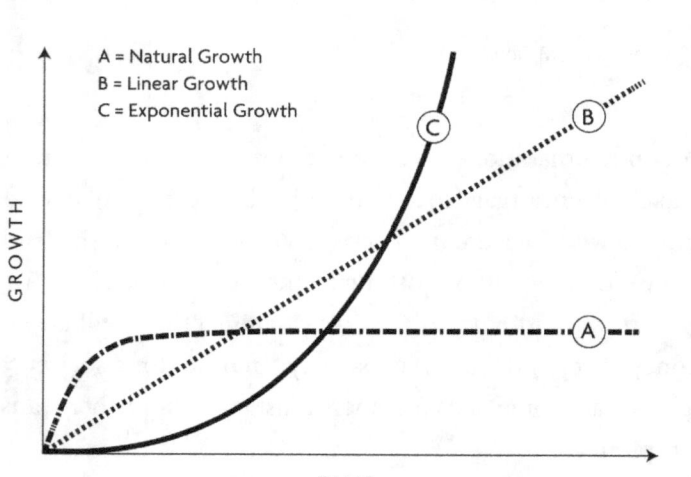

BASIC TYPES OF GROWTH PATTERNS   *4*

A = Natural Growth
B = Linear Growth
C = Exponential Growth

3. **Instability**

Compound interest leads to exponential increase in stored money (solid line).

GROWTH

TIME

As an example, consider the legend of the Persian king who accepted the gift of a precious chessboard. Delighted with this exquisite tribute, the king offered a reward in return, to which the bearer suggested the apparently modest deal of one grain of rice on the first square of the board, two on the second, four on the third, etc, to which the king agreed. When it came to counting the rice, however, the king realised that he had been tricked, as by the twentieth square he owed over half a million grains ...and by the last square on the board the king needed to find 18,446,744,073,709,551,616 grains, more than could be found in the entire world, to reimburse his "gift".[5]

Such abstract mathematical growth is obviously out of synchrony with the external biological realities it manipulates.

4. **Perpetual economic growth**

To service debt expansion the economy must grow indefinitely or else the system will collapse.

Central banks and the Business Cycle

Central banks have two main purposes (i) to function as a bank for other commercial banks (ii) to manage the rate of money creation; by creating reserves or by regulating the commercial banking system's ability to create money through control of the interest rate. This manipulation gives rise to what is known as the business cycle (the 'rowing' described above); at low interest rates the rate of lending and therefore money supply increases. Once inflation arises the banks raise interest rates, money supply drops and their level of profit from debt repayments increases. As both central and commercial banks are profit making institutions this cycle of high lending then contraction is integral to their functioning and must be maximised.[6]

"Whoever controls the volume of money in any country is absolute master of all industry and commerce"

– U.S. President James A. Garfield[7]

The dynamics of positive interest can therefore be seen as a powerful tool of hegemony. The perpetual transfer of wealth from net debtors to net lenders creates an undue concentration of wealth in one sector of society. The ability to manipulate money supply gives this same sector of society a means of influencing the behaviour of all others; agriculture, industry, commerce and government. This unprecedented power is amplified by its own invisibility, hidden from normal perception and scrutiny in the unconscious base assumptions of the public.

Healthy Money

With such dangerous and damaging social consequences it is perhaps not surprising in retrospect that positive interest money was considered immoral and illegal for millennia by the cultures of the three near eastern religions (the sin of usury). What then are the alternatives?

(i) Interest free money. Islamic banking remains a healthy interest free system and is alive and well today.

(ii) Negative interest money. That is, money which depreciates in value over time. Negative interest money systems have been used in the past and hold great promise for the future.

The Miracle of Worgl

Modern historical experiments with negative interest money originate from the work of Argentine-German economist Silvio Gesell (1862–1930). A businessman with firsthand experience of the volatilities of currency, Gesell sought to develop socially responsible capitalism. This position would earn him few friends in a world oscillating between the extremes of Marxism and Fascism and he died before any of his theories could be put into practice. However his legacy remains; indeed the most notable community currency experiment on record was based on Gesell's writings.

In 1932, during the dark days of the great depression, the Austrian town of Worgl was struggling badly with steep debts and high unemployment. In the face of this dire situation the mayor, who rejoiced under the title of Michael Unterguggenberger, persuaded the town council to enact an emergency currency along the lines of Gesell's theories. Worgl issued 20,000 local schillings which had to be validated each month with a stamp worth 1% of the nominal value of the note. In essence, Unterguggenberger had created money which could only be used locally and which contained an inbuilt hoarding tax (a negative interest rate).

The circulatory quickening effect of this new currency soon became apparent; taxes were reimbursed, streets rebuilt, the water system redeveloped, forests improved and even a new bridge was soon completed! Unemployment dropped in Worgl as it rose throughout the rest of the country, news of "the miracle" spread far and wide and around 200 other Austrian towns developed plans to emulate the Free Schilling. However, the whole enterprise was brought to a crashing halt in November 1933 when the issuing of currency was criminalised by the government under pressure from the Austrian central bank. Unemployment in Worgl returned to its previous levels and the town sank back into the Great Depression — the depression which would fuel the emotional background for the rise to power of Adolph Hitler.

Further Reading

Healthy Money, Healthy Planet by Deirdre Kent
The Future of Money by Bernard Lietaer

17/April

It wasn't just the ongoing emission of carbon dioxide or the ongoing destruction of ecosystems or even the ongoing depletion of the bedrock of our civilisation, fossil fuel, which was the source of my unease. It wasn't just the omnipresence of a flawed monetary system. The church experience had brought home the fact it was *everything*. I was now looking at the world in a different way.

It felt like I was waking up in *The Matrix*.

And all the while the Sydney environs were ever present; the very place where that Gnostic-inspired movie had been filmed.

19/April

I ended up queuing with an odd assortment of students and asylum seekers for interview — door to door leafleting was the job in question and I was given a backpack's full to start me off post interview. Seemed like a piece of piss.

It took me about ten hours of constant motion to post them all. Once back at the hostel I had to nurse my blistered and bleeding feet in the shower. I worked out how much I was owed for the stretch and my thoughts drifted back to the original advert:

"Break your back for a pittance"

20/April

Despite the blisters, getting some work had taken a weight off my mind and I started to relax into my situation more. I was getting a feel for the people in the hostel. It reminded me of those halcyon days as a fresher at University, though on this occasion I was on the outside looking in; disparate people meeting up, hooking up and hanging out. Letting themselves out for perhaps the first time in their lives. Along with the ongoing passers through there were a good number of people there for the long haul. They had formed into a lot of cliques like little families; some young gay Danes enjoying the bohemia of freedom in a foreign city, the English crew, the Irish crew, the Canadians, the Asians.

Everyone was in much the same boat, trying to have a good time but running out of money quickly. Such was the price of living in Sydney that almost everyone had to work. Those that did have a job spent their earnings almost immediately.

21/April

My feet were feeling the miles but I stuck at it. Thankfully, the office staff had miscalculated my load on that first job by a considerable margin and subsequent rounds were more realistically achievable. I was still barely breaking even at the end of the day, but some dollars was better than none dollars, and tramping the city streets was much more stimulating than hanging about the hostel.

Jason had left, to explore the fabled east coast, and my preoccupations were such that I was fairly oblivious as roommates came and went. But thinks changed when Mike breezed in. Looking very much like a lost member of The Strokes, he was a charming 20 year old German with an exceptionally casual manner. Mike was followed later in the day by Erin from the United States; a bleached blonde surfer-type whose whole demeanour said "U S of A".

22/April

At least that's what I had misheard his name as; 'Erin' was actually Ariel, and Ariel was actually an Israeli. One of those rare Israeli's that travelled on his own, and like most young travellers from that country he was taking some time out post military conscription service.

Ariel was a big talker, and we hadn't know each other long before he broke into a long confession about a prank he and some friends had just played whilst on *Koh Phan Ngan*. Apparently he and two other Israeli's had decided to play a game where they pretended to be three backpackers of different nationalities — Ariel playing the part of an American. Once they had got to know a bunch of people they brought up the concept of "Israeli travellers"; just to see what sort of response it would bring.

I'm sure that must have been an eye opener for them.

However, their little game got out of hand as the Island progressed towards Full Moon, they got to know more and more people, made more and more friends, got together with some girls, and so ended up leading their secret double lives full time for the entire duration of their stay.

Ariel had worked himself into quite a state of guilt over the whole issue and was preparing to come clean over email to his friends and former girlfriend from Sweden. He was obviously more sensitive than he looked and I wished him well with breaking the news.

23/April

Ariel came in from a wander around the area. He was pale, sweating and flustered. I wondered if he'd been introduced to the volley of verbal abuse that follows rejecting the requests of beggars around here. Turns out I was right. Welcome to Kings Cross.

Mike and Ariel had hit it off almost straight away. A yin and yang combo; Ariel completely conventional, a walking menswear catalogue, and Mike the indie kid who liked to be different. But they made a good team, and they enjoyed one another's company.

We three shopped together in preparation for our evening meal and I played the veterans role, showing them the sights. The city looks different when you've been out in wilderness. All those little glowing windows piled high in the night air, thousand upon thousand; portals representing lives. Ariel was remarking on how alien it looked after months in nature. I didn't know if *Hat Rin* counted as nature but I knew what he meant. It was good to realise that the re-culture shock wasn't just me.

24/April

It seemed strange, that here was I contemplating the nature of our monetary system and the gap between rich and poor, and here was I shuttling between the gated palaces of Bellevue Hill and the destitution and degradation of Kings Cross. A mere ten minutes walk the gap between different worlds.

Atop the eponymous hill were stunning views of the marine channel under blue skies. Looking in at the massive ornate houses and their gleaming cars whilst aware of the amount of money I was earning for this repetitive, physically strenuous task did nothing to improve my mood. For me though, this was just an illusion. I knew I could stick a flight on my credit card and probably be back at work the following week if I really wanted to. But I was aware of for how many people — countless people — in the world does this constitute life; daily physical exertion for barely enough rewards to make ends meet, all the while looking in on a world of relatively incredible wealth and privilege.

25/April

Another day in the midst of the "War on Terror". The War on War. That phrase is humanity in a nutshell — to solve war, you go to war with it.

The town centre shut down for a military band parade. I had a walk round and took in the spectacle. Families and children, all waving and cheering. An enormous war for oil and profit was going on, yet this display of militarism was seemingly the perfect day out?

I wondered what everyone else was making of this.

Did the children only see the colours and hear the music?

Did the adults only see an affirmation of identity, of shared purpose, of glory and honour, of past sacrifice and self defence?

I could only think of how difficult it was for one human being to kill another[1], a surprising reflection in light of our history, and of the massive amount of conditioning required to make this happen on demand through instructions from a command and control structure.

26/April

Manna. A job. A real proper job. After six weeks the agencies had finally come good and found me some contract office work. It was just office temping, but at $20 per hour was about as good as it could get for itinerant backpacker work. At two weeks long it was the perfect amount of time as well. Things were starting to come together.

While I was out tramping the streets, leafleting relentlessly, Mike and Ariel were having a good time touristing around the city. They had been on a day trip to the spectacular Blue Mountains out west and enthusiastically showed me the pictures on my return. It was a place I had wanted to visit but couldn't afford.

27/April

In my free time I was reading up as much as I could on mysticism. I had to try and make sense of my experience; to discover why other people did not know that it was the only real route to truth.

Trying to think of all the possible routes for development my culture could have taken...

Contrasting the different scriptures of the different world traditions...

Trying to work out why this knowledge had been hidden or lost...

...I realised I was facing up to a puzzle as monumental and as old as humankind.

28/April

The conventional mind never fails to beggar me. Ariel was looking for beach-wear and asking me if I knew any local outlets; but he didn't want just any-thing, he wanted:

"A *Brand*"

What really took the biscuit though was when he proffered his philosophy of life:

"Lifestyle. That's what it is all about. Lifestyle."

Speaking in reference to his parent's set; their houses and their yachts and their cars and their vacations. What he was aiming to have twenty years down the line.

The possibility engine of my mind almost broke down under the weight of the absurdities of the statement. *Lifestyle*? In *Israel*? In *twenty years time*?

I thought it'd be lucky if there was a crater where Israel used to be in 2026.

29/April

I congratulated myself for finding proper work with a night out with Mike and Ariel. I would never have expected that it would take me six weeks to go out and have a beer in Australia. But in truth I hadn't felt up to it before now.

Kings Cross turned out to have a vibrant nightlife as well as the sleaze — think London's Soho — and it was like being a student all over again. For the first time since I had arrived in Australia I socially relaxed. I forgot about everything that had been on my mind — life was good, girls were pretty and global warming definitely wasn't happening.

30/April

Mike had come to Australia to follow his ex-girlfriend Relinde, who had dumped him prior to going travelling. Strangely he had no intention of winning her back, he just wanted to hang out. He invited her round to the hostel to meet us. She was a nice girl, petite and intelligent and not the sort I would have visually tied in with lost bass player Mike. Relinde had been here for months and had seen and done a lot. She had taken the *The Ghan* across the Outback. She had been on a tour around Uluru with one of the few non-native tourist guides who had been initiated into Aboriginal culture.

Apparently this particular guide was something of a legend; he had spent years living in Arnhem Land with the natives and had earned their respect, and consequently many of their secrets. Relinde's description of their culture, of how alien it was to our own and consequently the difficulties the Aboriginal people had of integrating into our way of life, more than piqued my curiosity. For instance, they didn't think of themselves as people living on the land, they thought of themselves as:

"We, the Land"

Listening to her made up my mind; finances be damned, I would have to see the Great Red Centre before I left this place.

GAIA

"For me, the personal revelation of Gaia came quite suddenly — like a flash of enlightenment. I was in a small room on the top floor of a building at the Jet Propulsion Laboratory in Pasadena, California. It was the autumn of 1965...and I was talking with a colleague, Dian Hitchcock, about a paper we were preparing...It was at that moment that I glimpsed Gaia. An awesome thought came to me. The Earth's atmosphere was an extraordinary and unstable mixture of gases, yet I knew that it was constant in composition over quite long periods of time. Could it be that life on Earth not only made the atmosphere, but also regulated it — keeping it at a constant composition, and at a level favourable for organisms?"

– James Lovelock, 1991

It is fitting that in the 1960s, the decade which gave us for the first time those iconic images of the earth suspended in space, a new scientific vision of a whole earth was formed. The original hypothesis developed by James Lovelock in partnership with Lynn Margulis has now advanced to the extent that Gaia theory represents an inter-disciplinary framework for the scientific study of life on earth.

The Earth System behaves as a single, self-regulating system comprised of physical, chemical, biological and human components. The interactions and feedbacks between the component parts are complex and exhibit multi-scale temporal and spatial variability. The understanding of the natural dynamics of the Earth System has advanced greatly in recent years and provides a sound basis for evaluating the effects and consequences of human-driven change.

– Amsterdam Declaration

Gaia theory is the crowning achievement of a new trend in scientific thought which traces its origins from the beginning of the last century. Movements in diverse disciplines heralded the rise of systems thinking; the discovery that in the living world, parts can only be understood within the context of the wholes in which they take place.

- In biology, individual organisms could only be understood through comprehension of the relationships between their various components, and of the self-balancing feedback systems which enable self-regulation and homeostasis.
- In psychology, the discovery that perception related to patterns of organisation which were qualities of organised wholes and not isolated parts.
- Ecology provided systems theory with its central metaphor; that of the network. At all scales, life is found to be composed of networks nesting within other networks. An understanding of cyclical resource flows and community interaction was also taken from the study of ecosystems.
- In biochemistry, self-organisation, the duality of structure and function each giving rise to the other, became the hallmark of living systems.
- In mathematics, the discovery of non-linear equations enabled mathematical representations of the infinitely complex structures found in nature.

This new understanding departed most radically from convention in the field of physics, where scientists were confronted with the revelation that subatomic particles were not particles at all; they had no meaning as independently existing entities and could only be understood as a set of relationships that reached outwards to other things.

This synergetic combination of multiple specialties and myriad minds has allowed science to develop a holistic understanding of life on earth for the first time and, in Gaia, a grand metaphor which transcends pure rationality and the limits of science. At the beginning of the new millennium, a new world view has emerged where life can now be defined as a property of planets rather than individual living organisms.

Further Reading

The Hidden Connections by Frijtof Capra
Revenge of GAIA by James Lovelock

01/May

The job was in a geriatric hospital in the western suburbs, filling in as maternity cover for the payroll department. Sylvia was the manager, from Sri Lanka originally, Rick and Molly were Aussies. The work was straightforward and the people were lovely and the day flowed by pleasantly. I even got a free lunch as part of the deal.

02/May

Mike and Ariel were apples and oranges in terms of personality and interests but they had fun hanging out and chasing girls. I had to admit that I admired them both. They were completely fearless. Two young men with their whole lives ahead of them, filled with complete optimism and confidence. Having them around had really helped me scrape myself off the ceiling.

Mike, the dark horse, was currently winning on the girl front. He would slope off to watch an obscure band of a week night, one that no one else wanted to see, and then not return until the next day. I suppose it comes with looking like a lost member of a rock group.

03/May

The internet is a truly wonderful tool when you are far from home. I had been spending some time in the net cafes after work, and I had become engaged in a lively online debate, starting on corporations but progressing into a general cultural critique of the way things are currently done. It provided the perfect outlet for my developing worldview and in contributing to the discussion I realised that I had achieved a better perspective for conceptualising the global issues of the day than most. A book had been growing inside me since I had been away, the only real way I could think of expressing myself fully, and I now realised that I did have something to say.

The monotony of the office temping tasks gave me plenty of day dreaming time to compose my answers in the discussion. I knew that my 'opponents' in the debate were just me as I was before I travelled. To those living in a stable western society everything seems A.O.K., with the ongoing march of globalisation, westernisation, marketisation and liberalisation sure to raise the living standards of the world. It's only when you get out into the world itself, and experience the abstracts of global problems becoming real issues affecting real people that you realise how much your former opinion was helping to create the damage in its unthinking support of the *status quo*.

I was amazed during the online debate at how my propositions of causalities behind the problems seemed to ricochet almost automatically, bouncing from belief systems rather than being taken onboard as useful concepts that should at least be researched and discussed more fully.

How to elucidate the true picture of the world?

I sat on the bus on the way back from work, daydreaming out of the window as intensely as I could, trying to find a starting point for my argument. As I watched the people walk to and fro, going about their Sydney lives, I suddenly realised I had found it. What was real? What was I really seeing as I looked out the window? The people were real, the buildings and roads were real and the odd patches of trees and grass were real. But what about the reasons people went about their daily lives? Were the jobs and offices, institutions and government, police and civilians, and all the other labels people applied to themselves and each other and worked according to really real? Was the money they all worked for real? Or were these all just frameworks determined by agreement of thought?

I realised just how much of the immense drama of human life was played out in a framework of thought. But then, even the buildings and roads were direct products of human thought, as were the trees and grass that were allowed to exist in the city, as were the people themselves, shaped by parenting, culture, education — the thoughts of others ...and their own thoughts.

Of course, those invisible, intangible things that created the experience; thoughts and emotions, were as real as the solid objects I was perceiving.

Just as the ancients had taught, it is all a product of Mind.

I sat back in my seat, rather pleased with myself for achieving this metaphoric starting point. The bus turned a corner and a view of the Sydney skyline flashed up like a frame from *The Matrix*.

04/May

Almost the victim of an unprovoked assault. One large vagrant loomed at me for no reason out of the commuting crowds as I was on the way to work. Eyes staring madly, he cut across my path and raised his hands threateningly towards my throat. I stepped back with apologetic body language, hands up, trying to defuse a situation I had in no way initiated. He veered away at the last second, a flicker of sanity dawning in his eyes at the moment just before I had to physically defend myself. Adrenalin coursed through me and my mind boggled at the random insanity of the place.

I need to get out of here.

05/May

The work wasn't physically or mentally stimulating but I was enjoying my time there, mainly due to my co-workers. In former employment there had been times when I worked in offices overrun by bitchiness and the clashing of petty egos but in this place the atmosphere was lovely and relaxed, just like the people who worked there.

My *qigong* practice was taking place later in the evenings as I worked till five p.m. then had to commute back to town. I was treated to one of the natural wonders of this part of the world as twilight fell over the botanic gardens, stars winked on in the sky and the moon glowed bright. A line of dark shadows rose birdlike from the trees. A cloud composed of hundreds and hundreds of shadows ascended before radiating out in long columnar lines across the skies, parading to destinations unknown. Their flight was eerily silent, save the occasional cry which echoed in the stilling air. Silhouetted by moonlight, the Flying Foxes of Sydney were a magnificent sight to behold.

Back at the hostel, Mike had found new female company. I'd returned from work to find them in bed, but thankfully not *en flagrante*. A local goth chick, she was a fragile little character. Having her first experience of foreign sex, she was surprised and delighted to find that Germans were:

"like us"

And:

"not wogs"

06/May

My *qigong* practice was really beginning to reap rewards. I woke up infused with a sensation of pure joy. It felt great to be alive. I was unbelievably energized with the joy of living and really looking forward to the day ahead.

When I stood up from my bed it felt as if I was floating on air rather than standing on legs. As if my body didn't have to expend any energy to move about, rather gained energy from everything I did. I felt like a child again; I just wanted to go outside and play. I breezed down to the botanic gardens for my morning practice and all the while I just felt what I can only describe as The Joy of Walking. There was a kind of ecstasy filling my being, and such joy in the simplest of tasks. I didn't walk, I floated down to the park, and the sensation of every step of the way was utter bliss.

Even the fact I was walking through Kings Cross didn't affect me today. I could appreciate everything; even squalor looked and felt good.

07/May

I found Mike easy company, but Ariel on the other hand had a tendency to occasionally rub me the wrong way. Verbose and intelligent, he was a dominating social presence as well as being a physically powerful bodybuilder and ex-soldier. He was the sort of guy you shouldn't like; handsome, intelligent, talented, and knowing it. But for all his superiority and arrogance, somehow at the end of the day you couldn't help but like the guy. Underneath the carefully maintained facade he had a heart of gold. He really did care about people. His sociability came from a genuine place.

08/May

Not feeling the sharpest this Monday at work. I had been awoken during the night by the sounds of coitus in the dorm. What can you do? I tried to shut it out with headphones and go back to sleep. I didn't know whether to be annoyed or impressed. Thankfully I could do what was being asked of me at work in my sleep.

On the way home from work two old men bumped into each other in the midst of the busy commuting. They shouted curses at one another before squaring up and throwing punches. They ended up rolling about on the ground whilst the wave of people parted and formed a crowd around them. Various members of the crowd were imploring them to stop. Suddenly they did, comprehension dawning. They both looked extremely conciliatory as they were helped up, and apologised to one another before shuffling off in opposite directions, faces reddened and heads downcast.

Just another day in Kings Cross. I had lost count of how many fights I had seen or heard since I'd been here. I would sit in my room some evenings and take in the night air and the passing spectacle of the junction of the main road. Sooner or later something would kick off. It seemed to be catching in the environment; it wasn't just the daily screaming matches between the home-less and junkies, but fist fights between the drunken revellers that descended at the weekends.

09/May

Mike's amorous adventures meant that Ariel and I were spending more time hanging out together in the hostel. I noticed that for all his bravado and puis-sance, Ariel had a real problem with being alone. He couldn't cope. It was something he shared in common with the other Israeli travellers that I had observed on my journey and it reminded me of Plutarch's description of the Spartans;

"...did accustom his citizens so, that they neither would nor could live alone, but were in manner as men incorporated with one another, and were always in company together, as the bees be about their master bee."[2]

Israeli's as contemporary Lacedaemonians? It was a stimulating intellectual reflection. A modern day society engineered for warfare?

10/May

Mike and Ariel had departed whilst I was at work, journeying up the East Coast, but they had left a note wishing me good luck. I was touched. I realised it was ironic that here was me and my preoccupation with all things 'spiritual', whereas the only two people in the room who were really being spiritual were the ones just enjoying life, making friends and building bonds. Expressing themselves fully and honestly.

12/May

The work had been a dream. Making money, keeping busy and meeting great new people all at once. The small team had really made me feel welcome, and each day was a new revelation in the mutual discovery of personality. Rick and Molly were the first real Aussie's I'd got to know. Rick was a family man, happily settled in middle age, enjoying life in this beautiful part of the world. Molly was a younger woman, and she traced her ancestry back all the way to the very first prison ship that had disembarked in Sydney Cove. You can't get much more real Aussie than that, without being an Aborigine. Sylvia had had a varied international career before settling here with her family, which included degree studies in the then Soviet Russia, and there was a lot in common we could talk about. They wished me well on my travels.

13/May

I booked a ticket to Byron Bay following completion of the contracted work. Jason, whom I had met in my first week in the hostel, was back following his saunter up the east coast, wings prematurely clipped due to lack of cash. He had decided to fly home. I got a game of football in with him and the guys from the hostel, but even in the midst of leisure activity the sickness of the area was still in evidence.

A middle aged lady collapsed on her knees at the side of the football pitch and started crying, having an anguished conversation with thin air. I recognised her as one of the number of Kings Cross casualties that had become familiar. She crawled slowly across the pitch and we collectively ignored her, the play going on around her as if she wasn't there. She eventually made it across to a covered set of seats where some of her similarly mentally ill friends were hanging out. I was just thinking of how the episode typified the whole place when a fat passerby, seeing the eleven-a-side game of football going on, stripped naked and began running about the field.

Byron Bay

14/May

There are few things I enjoy more than travelling somewhere new. My spirits couldn't fail to be lifted by the open road and the brightness of the southern sky. I'd booked a hostel which was well rated on the internet and was pleasantly shocked with how nice it was, though it wouldn't have taken much to outshine the backpacker palaces of Kings Cross.

As I unpacked in my dorm I noticed a sticker on the side of the fridge with some chagrin:

"SAVE THE PLANET!"

It stated, along with a picture of said Globe. Underneath in small type:

"please turn off the lights when you leave the room"

After all my discoveries of the past year, all that I'd been researching and studying, the message hit me like a kick in the teeth. *"Save the Planet?"* This assumes that firstly, The Planet, the Entirety of Life as We Know It, is indeed in need of being "Saved", secondly that this will be achieved by switching off the lights when I'm not in the room, and thirdly, that sticky messages on the side of fridges are all the education required to realise the Salvation of Everything.

My head spun as I contrasted the societal response to clear and present dangers, such as hostile foreign military action, with the most important task that could realistically be conceived; "Saving the Planet". I'm sitting in a building basking in how many hours of bright sunshine per year? And they are telling me that switching off the lights will save the planet. Some solar panels on the roof of the building, energy efficient appliances and intelligent sensors/switching devices might be a better way to go about "Saving" things. In the short term, not money, thus creating the need for said sticker. Rather than moderately reducing the CO_2 emissions caused by the electrical lighting requirements of the building, and relying entirely on my whims to do this, why not just solve the problem altogether by switching to renewable sources completely? *Anger* surged.

I searched for the funny side of the situation as I attempted to let it go; I figure you can only have a sense of humour about these kinds of things, lest you lose your mind.

I took a wander through the hostel, which had an open plan design centred on an outdoor communal square complete with bar. There was a large indoor communal kitchen near the main entrance. The openness, cleanliness and freshness really felt great. I spent some time exploring the locale before heading back to make dinner, resolving to meet my fellow hostellers in the process.

It didn't take long to meet Jo, who was making a stir fry as I walked into the kitchen with my collection of foodstuffs. She was frying with ingredients including peanut butter. It was the perfect introduction to her unapologetic uniqueness. I liked her immediately.

We got talking and I ended up having dinner with her and some of the other hostellers. Jo was a 25 year old Canadian on a working holiday visa as part of a round the world trip. The bizarre stir fry was a good introduction to her character — completely unconventional and completely happy in her unconventionality. She was a lovely girl; clever, interesting, artistic and offbeat and it was a pleasure to make her acquaintance. Such is the nature of Byron Bay hostels that there was a happy hour on, and the night progressed from sitting round a table, eating, to sitting in the main patio, which doubled as a bar, drinking. The place got pretty crowded with young travellers and the night progressed. Everyone ended up going to the clubbing staple of Byron Bay, tables were danced on and much alcohol was consumed. I lost track of Jo at some point, but met a lot of random people along the way...

15/May

I tried to make sense of my shattered memory. It had been a fun evening but romantic interaction with delectable girlies, which I always feel is the barometer of the worthiness of a hangover, was sadly lacking from my fragmented memory. My wallet was also feeling the strain of the revelry.

I met up with Jo again over lunch and we decided to go for a walk up to the point overlooking the Bay. The beach was spectacular; golden sands and gently foaming breakers all under a bright sky scattered with fair weather clouds. We got to know each other as we walked along, across the beach and through groves of temperate coastal rainforest. We climbed the path up the hill to the lighthouse, past scores of surfers paddling lazily behind the point, awaiting the next big break. The sea, deep blue and white flecked in the distance, was crystal clear close to shore, and Jo gasped as she caught sight of dolphins cresting speedily through the waves, revelling in the experience of the sea as were the people.

As we reached the top of the hill and the lighthouse, seabirds soared overhead on the updraft from the cliffs, and men in hang gliders hovered in no less an impressive feat. It was a spectacular image. The view from the peak was equally astounding; to the south breaking waves, white sands and green forest and fields stretched endlessly into the distance, to the north and west the pleasant town itself nestled in the bay, surrounded by lush veridian fields and pasture which stretched towards the horizon, a horizon ringed by forested hills. It really was magnificent, and I could only think of how much a paradise this part of Australia must have seemed to the generations that had migrated here from home. With the lighthouse overlooking the soft emerald bay, it was like Great Britain was dreaming a dream of itself, as if a part of the British coastline had been pulled down towards the tropics so that people and dolphins could frolic in the sun and warm waters.

It was wonderful to share the experience with someone like-minded. It was the first time since I had arrived in Australia that I had found another backpacker that I could talk to. Really talk to. I felt that I could talk to her about anything. I confided in her my discontent with the endlessly repetitive hostel drinking scene. I knew I would have felt different had I been eighteen but I was eighteen no longer. She was equally unimpressed with the constant drinking and wanted to get off the beaten track; camping out, exploring national parks and WWOOFing her way around Australia. Whereas I was trying to straddle two worlds — the alternative and the touristic — keeping one foot in both, she was completely comfortable in her own identity.

We checked out the little museum in the lighthouse and looked out at the sea, as hang gliders and birds of prey coasted on the updraft from the headland. A wonderful place.

16/May

A day of just hanging out. It was an easy thing to do in a place where the weather was so good. Jo and I talked about one of the shared interest we had happened upon the first time we met; WWOOFing. I had noticed a lot of good projects going on in town and wanted to find a place here before moving on to Brisbane to look for more work. Jo on the other hand wanted to explore the national parks further north, before finding a place to WWOOF.

I contacted the first of the hosts I'd highlighted in my book. He didn't seem too enthused, and let me know he already had enough WWOOFers on board at the moment. He told me to call him back at the weekend, as by then he would have a better idea of his upcoming labour needs. I tried the other entries I'd highlighted and they were either not currently looking for workers or impossible to contact.

17/May

I took some surfing lessons while I had the chance. A childhood dream fulfilled. I sat beside Jo's German friend, Magnilda, on the surf school bus. At introduction time I found out that she, like the rest of the noveau surfers, was around eight years younger than me. Since I'd been away I had always been so impressed with people who had decided to go travelling on their own at such a young age. I would never have had the courage.

We had one of the usual backpacker conversations about working holiday jobs here in Australia and she made an interesting observation on the art of sales. Magnilda had quit working as a salesperson for a trendy sunglasses company as even though she was good at selling, she did not like the very concept of making others unhappy, of making them feel like they were lacking something in their lives. It was an interesting philosophical point from a pretty young girl who on the outside looked the epitome of convention.

The surf lesson was amazing fun. Today we were only learning the initial phase of standing up on the board on the back of small breaking shoreward waves. It was plenty challenging but a lot easier than other balance-requiring activities I'd tried in the past like ice or roller skating. There were several crashes though, and I bumped into Magnilda a few times, purely due to a lack of skill and not in a prepubescent attempt at flirtation (a methodology I've outgrown, just), and was surprised by her volcanic temper; another facet of her personality that could not be guessed at from her exterior.

One of the two surf instructors, an 18 year old local, took his board further out into the point break proper and dazzled us all with a display of wave riding virtuosity. Managing to stand up on the board is one thing; actually surfing inside the curl of the breaking wave is another, requiring an incredible amount of dynamic skill and balance. I smiled wryly as I watched him attempt to engage Magnilda following his performance.

That evening there was a surf night, which consisted of basically a background video of some impressive wave riding on a large screen in the pub of another hostel, and we all attended. Later I made sure I caught the European football in the pub with Dan, the other Scotsman on the surf course. I got to see my all time hero, Henrik Larsson, come on as a substitute and change the course of the European Cup Final in Barcelona's favour with his trademark understated selfless genius. Glorious.

18/May

It was time to say goodbye to Magnilda and Jo, who were travelling up the coast with a young Canadian friend who owned a car. Jo and I's WWOOF plans were currently divergent, but we hoped we'd get an opportunity to WWOOF together in Brisbane or further north. As we parted company I chuckled at the tale of Magnilda's perspective on the prior evening. Seems surfer boy had failed miserably in his amorous attempts, and I empathised with his plight, as I had thought he had been doing so well.

Day two of surf school was shared with the three Israeli travellers that had been roommates of Jo, as well as Dan from Scotland. Dan was quite a bit older than me and was enjoying a short break in Australia from a working holiday in New Zealand. He was also a WWOOFer and we shared the usual "what do you think of WWOOFing?" chat. Some of the more out-there entries were always a topic of amusement.

After the course the surf instructor enlightened us of one of the back-packer delights of Byron Bay — a weekly charity kitchen put on by a local church. We went along and joined in the extremely mixed crowd taking part; backpackers, students, missionaries, little old ladies, nutters. It was an interesting combination, and we ended up in the queue beside one homeless guy who was taking issue with the speed of proceedings with the mixture of loud voice, lack of social constraints and warped humour only found in the panhandler. He especially had issues with all the;

"Backpackers with Credit Cards!"

Benefitting from the;

"Free Feed!"

A young Seventh Day Adventist missionary engaged us with the forced friendliness of the zealot, and his cadre of Vanuatuan convertees introduced themselves shyly. The food wasn't half bad and the gathering turned out to be a lot of fun.

19/May

At $60 for a two day lesson, the surfing had just about killed my finances for the week. I spent most of the day wandering around town and trying not to spend money. Resting up at the hostel, I browsed a copy of *The Hitchhikers' Guide to the Galaxy* that Mike had given me. I thought back to when I had read about it first time around, as a twelve year old schoolboy. Back then, the anti-conventional irreverence that ran through the book had irritated my sensibilities. Now, I understood the traveller's perspective of knowing that there was so much more out there, and wondering why so many accepted mundanity for their whole lives in comparison.

An east coast surf tour — think 18–30 — came through in the afternoon.

Loud rock music was played over the hostel speaker system till 4 a.m. It was definitely time to leave for Brisbane, but I thought I'd hold on for one more day in case the WWOOF I'd been planning on became available.

20/May

I contacted the WWOOF host again. He wouldn't be able to take me on. Time to cut my losses; luckily I had an open ticket to Brisbane, and I could jump on the first bus on Monday, and begin the attempt-to-find-work merry-go-round once more.

21/May

The past few days had seen the usual hostel comings and goings. Two English girls had spent an entire day zonked out in bed following a trip to Nimbin, the local hippy paradise, a psychotropic wonderland on open sale. One of them gave me some good tips on life in Brisbane. She had spent some time there working as a stripper, and consequently had less money problems that I did. I met a guy from Newfoundland, a place I could never have previously imagined people actually living. He was brand new.

22/May

Early morning practice on the beach, looking out to sea. It was perfect. As I finished my meditation and slowly opened my eyes I was treated to an incredible sight; a dolphin porpoised rapidly across the breaking waves, more powerfully and gracefully than any surfer, before diving down and then leaping up in a spinning crescent through the air. A spontaneous expression of pure joy. Farther out, in the deepening waters of the bay, a foaming wake preceded the roll of a huge body and the splash of a great fin. It was a humpback whale. What a way to leave.

Brisbane

22/May

The hostel was located in a suburban cul-de-sac not far from the centre of town. A shuttle bus was provided which I shared with some girls from Korea.

It was a functional place and I introduced myself to my roommates. One of them was an Asian man whose first words to me post name exchange were:

"I ...don't...like...Australia"

You and me both, mate.

His name was Ki Duk, from Korea, another post-military service traveller. Australia was a popular destination for young Koreans for the twin purposes of learning English and making money to fund their studies, and this was the case with Ki Duk. He had quickly discovered that English wasn't to his liking and that left money making as his primary concern.

23/May

I got orientated. Brisbane seemed a nice place — big enough to be interesting but small enough to get around. The centre of town was compact, though the planning muddled. A few impressive skyscrapers stood like sentinels along the banks of the Muriwara, the muddy river that meandered lazily through the centre of the city. There was a recruitment agency high on one of the floors of the biggest tower, and I got a chance to appreciate the outstanding views over the surrounds as I went about my job searching.

I figured it had taken me six weeks to find work in Sydney, so I wasn't going to panic if nothing came up in the first few weeks here in Brisbane. I looked into the possibilities for harvest work in the surrounding area as well. I found that the government provided *Harvest Hotline* was next to useless and got similar feedback from the other travellers to whom I mentioned it.

24/May

Ki Duk and I did some city exploring together. A river taxi took us from the suburban hostel into the centre of town. The riverside held a clutch of impressive new developments, and we took in the Queensland Museum. The natural history of the region was a litany of disaster, that familiar colonial tale, which ended on what the museum authorities must have thought was the bright note of the survival of the Northern Hairy-Nosed Wombat. The few remaining specimens of this sorry little creature now existed in only a singular forest in central Queensland. Across the grand hall of the museum, directly juxtaposed to the survival of the little wombat, and WITH NO SENSE OF IRONY, was the story of the coal wealth of Queensland, and its projected exploitation into the far future as the economic basis of the state.

25/May

The hostel was quiet, with a 10 p.m. lights out rule. After Byron Bay's rock music to 4 a.m., it seemed that I'd switched from one extreme to the other. With not much else to do, I spent time reading a copy of the *Bhagavad Gita* I had picked up in Byron Bay. My mind lit up as I took in the stanzas. Even though I was trained in a system of Buddhist origins, I recognized the universal nature of the methodology and the message through the haze of time and culture.

One stanza in particular resonated as I considered the state of the earth today:

> *Demonic men cannot comprehend*
> *activity and rest;*
> *there exists no clarity,*
> *no morality, no truth in them*
> *They say that the world*
> *Has no truth, no basis, no god,*
> *That no power of mutual dependence*

Is its cause, but only desire.
Mired in this view, lost to themselves
with their meagre understanding,
These fiends contrive terrible acts
To destroy the world. [1]

26/May

As I went through my morning *qigong* practice in the local park I was surprised to find all my power had gone. I practiced my routine as usual, but
rather than swelling outwards and moving my body gently in *qi* flow, all of my
flowing energy had gone. I tried to work out what was happening. It was as if
my *qi* had decided to target something specific deep within my body. A suitable analogy for *qigong* is that it is like running your body on a higher grade
of oil — everything runs smoothly but life still happens. The extraordinary
benefits are cumulative over years of practice.

My suspicions were confirmed as I began to feel progressively sicker as
the day wore on, with a building pain in my abdomen. My thoughts strayed
to the previous day's chicken dinner. By the evening I was in full blown food
poisoning; sweating, feverish, leaking out of one end. The pain in the pit of my
abdomen was excruciating.

I can tell you that there are not many experiences more unpleasant than
getting sick, by yourself, in a foreign country.

But a guardian angel arrived in the form of Ki Duk. He was instantly
concerned with my well being and went out of his way to help me. He had a
little DIY acupuncture kit with him and I let him have a go at sticking needles
in my hand. Some Korean medicine from a local store got me leaking at both
ends.

28/May

I had spent the past two days in complete agony, shuffling between bed
and toilet but through it all I was touched by the incredible kindness and
thoughtfulness of my new friend. He had been there for me when I needed

help the most. Once my appetite returned he even made me Korean seaweed soup.

"I'm... not...gay!"

He assured me with a laugh, as he considered all the TLC he was providing.

29/May

Jo had been enjoying herself immensely WWOOFing around northern New
South Wales, an incident with Chilli processing aside — after her description
of "like getting alcohol in a cut, but your hand is skinless and being doused in
alcohol with no numbing or relief", I made a mental note to wear gloves if ever
harvesting said fruit!

We were keeping in touch by email and provisionally looking into
WWOOFing together. There was a Yoga centre I wanted to stay at here in
Brisbane, but the monastic regime with ten hours of silence, minimum one
week stay and vegan only food didn't appeal to Jo. But she'd be hostelling in
Brisbane soon so there was a good chance we'd be able to hang out.

I considered my options as I went about my agency rounds. Money would
not last so I might as well stay in the city for a while and look for a job, and
hope that some office temping or the equivalent would come up, like it had in
Sydney. I was tiring of the relentlessly revolving door of hostel life and craved
some stability in my relationships. Moving into a shared flat might be the way
to go? Other travellers had recommended it over living in hostels. It'd be good
to get my own space. And it'd be cool to hang out in this city. It was the perfect
time for hanging out as well; the World Cup was soon to start.

30/May

Another unexpected ray of light from Ki Duk. How about sharing a flat with
him and some other Koreans? He tried to sell it to me as best he could; he
pitched it as a cultural exchange; I'd get a great apartment for a good price and
they would all (except him!) be keen to learn English. I wasn't too sure about
the idea of being the only English speaker amongst a group of foreigners but
he persuaded me to at least take a look at the place.

He introduced me to his friend Hwani, who was also interested in the shared condo. Hwani was a big bashful chap with a shock of thick black hair, this first conversation did nothing to dispel my fears over the probable English situation of the prospective household.

31/May

I got a phone call from Jo, who was WWOOFing somewhere in Queensland. The tension in her voice was immediately noticeable, and I was shocked as she broke down in tears during the course of the call. She was WWOOFing on a remote farm and felt violated by the treatment she was receiving at the hands of the father of the house. She had been expected to work for six hours on some household DIY task and then to make dinner and clean up after the family on top of all that.

I told her that what was going on was unacceptable and to get out of there as fast as she could. She responded that there was no public transport and that she was dependant on the family for mobility. Her credit ran out, the call ended and I felt bad for my friend. It was clear that she wasn't in any physical danger, just that she had obviously been exploited by the people she was staying with. It seemed she had encountered the dark side of WWOOFing; that nightmare possibility: alone, in a foreign continent, in a remote rural area, in a single farmstead, with no means of transport.

Alone, that is, apart from the company of an abuser and his dependants.

Little Korea

01/June

The apartment was fantastic. A new build in a brand new development integrated with newly landscaped parkland, situated only a few hundred yards from the centre of town. Three bedrooms, one ensuite, a bathroom and a large open plan living room with kitchenette. Downstairs there was a free gym and a free swimming pool, and outside the magnificently designed parkland with waterfalls, lakes and barbeque machines literally on our doorstep. Once the rent had been split between us, the weekly price worked out at less than hostel rates. Talk about landing on our feet.

After thirteen months on the road it was great to unpack properly for the first time. To relax and settle down a bit. I had only met Hwani, Ki Duk's friend and my new sharemate, once before in the hostel and I had been worrying about communication. However, like many East Asians his English was scrambled on first meeting due to nervousness. As we got to know one another throughout the day the situation improved and I started to get a picture of the personality of this big chap; another beautiful human being.

With some relief I took a phone call from Jo. She had arrived in Brissie and we made plans to meet up. She had bravely managed to stand up to the situation with her hosts, making it clear it was totally unacceptable, and had made arrangements to leave at the first opportunity.

02/June

Jo was thankfully was none the worse for wear after her WWOOF from hell. She was in the process of making a complaint to the WWOOF organisation about the experience. I met up with her and some new friends she had made at her hostel and we went for a walk round the city centre.

Brisbane sometimes gets a bad rap, but I found it a nice place to spend time. The weather was always perfect. Not too big, not too small, international class facilities and a gateway to the natural wonders of Queensland. The part about being a gateway to the natural wonders of Queensland only a hypothetical for me and my means unfortunately, but I could definitely see the potential of living and working here. It was a shame there did not appear to be much available going by the feedback from the agencies.

We walked around the riverfront, including the remarkable Brisbane artificial beach, and had lunch in the Botanic gardens, where we were attacked by a Kookaburra. Not something I had hitherto imagined would happen in my life.

Jo's friend Bernadette had recently arrived in Australia by herself and was experiencing that overwhelming sense of pressure that I had encountered when I first arrived. *No friends, no family, no job, what is my future?* Again, it turns out it wasn't just me. I assured her that things would fall into place. Bernadette had already found a great person to become friends with in Jo.

Back at the apartment I met our first new arrival; a shy young Korean girl who would go on to be known in legend only as "Casino Girl". Ki Duk had found the apartment through a contact he had back home, and it was in his interests to fill it up as much as possible. He had a plan to have seven sharemates; two per bedroom plus himself sleeping in the main room. He wasn't one to waste time and had been networking furiously over the past few days via the various Korean ex-pat services in town.

03/June

Casino girl would be moving out on Monday. She had been to the casino the night before. She had lost $5000. It had taken her six months of hard farm work, which she had hated, to make the money, and now she was -$500. I didn't know whether to feel for her or to be appalled. The girl obviously had issues; shy, with a defeated posture and terrible skin. But now she had a whole new issue. She was going back to the farm to try and build it up all over again.

04/June

Two new friends arrived in the shape of Kim Ji Eun and Lee Sung Hoon. Ji Eun was a cute, bright and friendly young girl, and Sung Hoon or "Bob" as he liked to be known in English, was a short, bright and studious young man. They had known each other from school back home in Korea, and were taking a year out to learn study and learn English here. Thankfully for the social sphere of the house, and my part in it, they could already speak English pretty fluently, and were keen to learn more. Cheeky chappy Ki Duk wasted no time in trying to mix it with me and Ji Eun.

05/June

A pleasant way to spend a winter's morn; stepping out to 20 degrees heat under Brisbane's blue skies. The library was an enjoyable fifteen minutes walk through the town and over the river. The facilities were fantastic, and better still were the quality of reference materials therein. It was simply ideal and I immersed myself in research for my book.

My Korean friends had all signed up on English courses and so the rhythm of the apartment felt very much like student life. In the morning I went off to the library, and they to their classes. We caught up in the early evening, and the guys were very generous with their food; keen to introduce me to the delights of their culture over dinner. *Doenjang jjigae*, a soup made from bean curd and tofu, was savoury and delicious. A rice cake covered in a spicy orange sauce was tangy and tasty, and also amusingly sounded like my friends name; "Duk" *(tteok)*. The irrepressible *Kimchi*, Korea's national dish, always featured; a fermented condiment made from cabbage with garlic and spices. My friends were over the moon when I said I liked it. Not liking *Kimchi* was like not liking Korea. Rice was rice, but in general there was a great range of flavour to the dishes. Ji Eun got frustrated trying to express how things tasted in English; apparently our vocabulary is too limited to express the subtleties of flavour and the experience of food to the Korean palate. We have 'nice', 'good', 'delicious' as positives ...and that's about it. Food to Koreans was like snow to eskimos.

Not only food, new words were on the menu, and everyone enthusiastically encouraged my first steps in Korean. I was praised for anything I got right. I thought this was funny as they were all so down on their own semi-fluent English while I could only pronounce and understand about five words of their language. In Korean culture people refer to their close friends with a family term suffix. They were all younger than me, so they were *dongseng*; Ki Duk *donseng*, Hwani *dongseng*, Bob *dongseng*, Ji Eun *dongseng*.

I was no longer "Andy", but "Andy *hyung*".

Elder brother Andy. I quite liked the sound of it. I guess I appreciated the implicit sense of intimacy and support that went with it. They all found the sight and sound of a westerner calling people *dongseng* more than amusing.

06/June

I was told I could greet Ji Eun with something that sounded like:

"Sa lang heh, Ji Eun"

Saranghe, Ji Eun

But I was wise to the ploy. She really was a delightful girl. A self-described "Korean princess". Most Korean women stay with their parents until they are married, the usual result of which is that they are spoiled by their fathers until well into their twenties. "Daddy's girls" we would call them. Ji Eun rejoiced in her sense of entitlement and she was loads of fun.

Today we were comparing blood types. It is an East Asian craze to read all sorts of personality details into someone's blood factors. Seems pretty weird. Apparently being type B meant that I was a womaniser and therefore bad news for the ladies!

I was starting to get a feel for the personalities in the house more and more. Ki Duk, tall, bespectacled and outgoing was the leader. Hwani, similarly bespectacled, was taller and bigger than Ki Duk but more circumspect and laid back, with a cheeky sense of humour. Ji Eun thought him the double of a famous Korean comedian, and with his bashful demeanour, random cheekiness, large size and unruly bush of black hair, I could certainly imagine it. Bob was one of life's brainiacs and Ji Eun a bubbly ray of sunshine.

Everyone was surprised with Ki Duk and I's ability to communicate despite his lack of English. I'd get what he was trying to say and he'd nod at me:

"Telepathic!"

It wasn't telepathy. One of the most surprising discoveries of my journey was the basic universality of human communication. Gesture, eye contact and body language could get you through the basics of life anywhere.

07/June

Ji Eun had only just moved in but she had already organised a party for her language course classmates at the house, and we were all invited. The guys were really making a go of it, and had bought loads of stuff at the Asian grocers for the meal. I watched them prepare it with no little curiosity. They had bought a lot of pork chops which they cut up with scissors into cubes. A wonderful sauce was prepared from a mix of bean curd (*doenjang*) and chilli paste (*gochujang*); this was *ssamjang*.

We had some soup at the house before going down to the park to cook the barbeque. That was another amazing factor in Australian life; the public barbeque machines. Something similar would be vandalised straight away at home. Maybe it was the weather? I pondered. It seemed to make people happier. Or maybe people were just anti-social back where I came from?

The highlight of the meal was the *ssam*. Hwani cooked the pork cubes on the Barbie. Ki Duk showed me how it was done; you take the barbecued meat, place it in a leaf of lettuce, add rice, a little bit of garlic, and a helping of the *ssamjang* sauce; sweet and spicy, then wrap it up. This was *ssam*. Then you pop it all in your mouth at once. It was absolutely delicious.

'Party' was the Korean term, but it was pretty civilised stuff. I'd used the term 'dinner'. It was more about the opportunity to socialise and eat together than an attempt to get blind drunk. I got to meet Hwani's girlfriend; a nice Korean girl I'd seen him talking to the other week at the hostel. I had a great time meeting students from all around the world; all in all it a lovely way to spend an evening.

08/June

At my lunch break away from the library I found I wasn't alone at this time of day like usual. I returned to find Ki Duk and Hwani draping a large Korean flag over the new wide screen TV they had just bought, along with a playstation. They had bailed out of their language courses four days in! Businessman Ki Duk had bigger fish to fry; he explained the kit was an investment to attract new tenants. He had also purchased the complete season one of *The O.C.* on DVD. The perfect pad had just gotten better; playstation and TV, just in time for the World Cup.

Seeing how much I was enjoying their cuisine, the guys invited me into their food sharing arrangement. I had nominally been making my own meals but it made sense after all they had been sharing with me. Now I'd chip in for the group shop, and just buy breakfast and packed lunch for myself. It worked out great; even including rent I was still spending less per week than I would if I was staying in a hostel, and now I could always join in with the utterly fantastic food.

The bean and chilli paste, *gochujang,* had blown my culinary world and with no little surprise that I had liked it, they were keen to feed me the many spicy delicacies of their cultural cuisine. The delicious bean and pepper paste featured in a lot of their cooking, and made everything taste good. I particularly liked *bibimbap*; a quick to make savoury snack that consisted of rice, mixed vegetables and *gochuchang*, with a fried egg on top.

09/June

Friday night was here and it was time for another party. By the time Ji Eun was ready to go out she was almost unrecognisable underneath copious quantities of make-up and wrapped head to toe in electric pink and gold. She was hitting the town with a female friend, and Bob was tagging along. The boys weren't clubbing types and they had decided to stay in. I joined them and we made the most of another outdoor Barbie. As a weekend treat, the guys had me toasting *soju* along with the meal. *Soju* was a fermented rice drink, a spirit rather than a wine; colourless and much like vodka without the aftertaste; extremely sharp but pleasant, it went down easily.

The correct way to drink *soju* was to down it in a shot glass after a wrap of *ssam*. It was incredible. The wrap filled with *gojuchang* a fantastic fusion of flavours; savoury and sweet and with a building fire from the chilli. A fire which was instantly quenched by the refreshing shock of the *soju* spirit; crisp and cool. Such contrasting sensations. A wrap of *ssam*; a toasting shot of *soju*; a wrap of *ssam*; a toasting shot of *soju*. The enjoyment of one added to the expectation of the other. The expectation of one added to the enjoyment of the other. It was culinary bliss. The object wasn't to get drunk; more to oil the wheels of the socialising …but you could easily lose yourself amongst the all the shots that went with the savouring.

Afterwards we shared some beers and talked about the football. The excitement for next week's South Korea match was building. The boys introduced me to Korean cards. Played with mini plastic cards with pretty oriental designs on them it was nothing like the usual ace of spades stuff. They were more like tiles than cards, and the guys would crash them down on top of each other with some force when they won a point. That seemed to be part of the fun, and they each had their own expressive style of clacking the cards violently down, one upon the other. They were anxious to get me involved and explain the rules to me, but I was exceedingly slow to pick them up, and eventually frustration got the better of them, and they advised me, as kindly as they could, to:

"just watch"

Bob and Ji Eun phoned us at midnight. Ki Duk asked if I wanted to go with them to a 'music room'? What was a music room? I asked. It was karaoke basically. I told them I didn't sing, but I was up for just going out. We walked into town and met up with the others. I was introduced to Ji Eun's friend; she insisted on going by the English name Alicia. She was tall, taller than me, and had a kind of Japanese beauty. Pretty and elegant with long, long luxuriant black hair that reached right down to her ass. I like that. Long legs. Lovely figure. She was also completely outgoing and friendly. I was entranced. I spent the night simultaneously trying to play it cool and make a good impression. The karaoke was great fun, worth even all the badgering at me to sing, and a youth of practice meant that they were all tremendous singers.

10/June

The parties were coming thick and fast. We had another sharemate staying over, a girl named Ha Eun, whom I had first crossed paths with on the hostel pick up bus on my first day in Brisbane. Ki Duk had let her stay over for a few days before her flight home to Korea. Ha Eun had quite a story; she was a student nurse who had been sent to Australia through an agreement between Australian government agencies and a Korean nursing agency. There was a whole group of Korean nurses who had come to Australia under this placement only to find out when they arrived that there was a legal loophole in Australian law that meant that they did not have to be paid any wages. What with being student nurses from Korea, many of them did not have the finance to return home (flights back were prepaid and prebooked as part of the deal) and had no option but to complete their work experience. They were effectively stranded, and the only avenue they had to make money was to work extra nursing shifts on top of their placement hours. The nurses had quite rightly kicked up a fuss about their predicament and it had caused a mini-scandal back home. Their rebellion had ensured the situation would not arise again, but there was no financial aid for Ha Eun and her colleagues stranded in Oz. I couldn't believe it when I heard it from her at first — modern day slavery in the first world.

As well as Ha Eun's leaving, a new girl had arrived and we held a little get together in our usual tradition. I had noticed one day during the week that Ki Duk had suddenly began taking very good care of himself; wearing his best clothes, spending a lot of time and attention on his personal grooming. Hot date lined up? I wondered. I found out what was motivating him when I was introduced to our new flatmate, a quite stunning girl called Ga Hee. Ki Duk had described her earlier in the day as beautiful but I'm not sure that word was an adequate description. I was completely knocked out by her as she sat and joined in our meal. She had more than the photogenic perfection of a cover girl; the fluidity of her movements and the intelligence in her eyes could not be held in any kind of stillness. She had such grace. Her body language was completely delicate and shy; the whole living picture of her simply the most erotic thing I had ever set eyes upon.

I left the party early, completely awestruck at the young woman I had just witnessed, to catch up with Jo at a bar for the England game. I felt pretty good to be leaving early; I hoped the new girl would notice my departure and that it would lend me an air of mystery and cool. There were a lot of English at the bar, including the folk we were watching it with, so I kept my Caledonian sentiments to myself. When I got back the party was still going on at the house. There was a massive and rowdy *clak-clak* game of cards going on, and I sat with the few who weren't playing.

The course of young love hadn't been running smoothly for Hwani, and he had just broken up earlier in the evening with his fledgling girlfriend from the hostel. He was consoling himself with alcohol. I kept him company. He introduced me to a Korean fermented rice beverage; sweet, milky white and mild of taste, and poured me some more *soju*. I was already buzzing from all the alcohol of the evening, and the extra *soju* was enough to tip me over the edge. Inhibitions blitzed, I started querying Ha Eun on the topic I was most curious about: Korean sexuality. I had always found that 'love hotel' culture of Japan mystifying, and I discovered they had the same things in Korea. Ha Eun also told me about the 'secret rooms'; hidden away in public buildings like shopping malls and office blocks; apparently most young men and women had their early sexual experiences, from loss of virginity onwards, in these rent-by-the-hour rooms, nominally intended for karaoke or video viewing purposes. Ha Eun said unwanted pregnancy was a big problem in their society.

"Condoms?" I wondered

"Men don't like them."

"You don't need to tell me that: I'm a man! Surely better than AIDS though?"

"It's a big problem in our society", Ha Eun repeated.

I had heard the same story in almost every country I'd visited. Even a few days earlier from Jo's hostel buddies in a similar conversation under a similarly alcoholic pall. Young girls everywhere were scared to tell their partners to wear one. Pregnancy aside, Russian roulette seems to be everyone's favourite game.

11/June

Ga Hee proved shy and only spoke Korean, though I was pretty certain that she could speak English, so over breakfast I tried to engage in her conversation, to break the ice in the living space as much as anything. Just simple getting to know you stuff; "are you studying just now?" and the like.

But I was met with a complete snub.

I'd never had that happen to me before; to talk to someone in a public conversation only for them to pointedly ignore what I was saying and then get up and walk away.

I was surprised and hurt.

"What just happened there?"

Ji Eun and Hwani explained that Ga Hee had felt uncomfortable with my degree of eye contact when speaking to her. Cross cultural protocol can be a tricky thing, especially when the gap is as large as between Scotland and Korea, and I made sense of the situation by thinking back to the last group of young Koreans I had met, where the men had spoke to me but the girls had quite liter-ally *ran away* when I tried to speak to them.

But for the rest of the day Ga Hee comprehensibly socially disempowered me, taking control of all conversations and social situations in the house. Any English conversation I tried to initiate with the group at large was immediately turned into a Korean one.

I had to take some time out to think about things. I went for a walk around the park. The experience had left me emotionally hurt and in the solitude of the park I broke down and cried. I was disappointed with my own weakness. Why was a crying? It was because I had been humiliated in front of my new friends.

I tried to get a grip of the situation. I didn't think I had done anything remotely wrong but the discomfort must obviously have been real in her head. I must have crossed some boundary.

With a wrench of self-awareness, I tried to see the world from her point of view. What would it be like to be a young girl, away from home in a foreign country, a foreign culture, for the first time? I'm sure the last thing she would have wanted would be to be stuck in a house with some random foreign guy leching after her.

I figured I should be more careful in future, the attraction to her would simply never go away but I shouldn't let it damage the situation in the house. A rather unnatural thought: perhaps if I managed to keep sex out of mind I might even become friends with the girl?

12/June

I worried about how things would pan out after yesterday's public snub, but over breakfast luminous Ga Hee looked happy and was smiling away.

She was in a great mood all day, friendly and bright and joining in the English conversations.

As we were all watching the world cup game she patted the space beside her on the sofa, to invite me to sit beside her. Talk about messing with my head. Was she trying to make reparations for her bitchiness of the previous day? After two days of pretending not to speak English, talking to me in her perfectly good English wasn't the way to go about getting back into my good books.

I let it go though.

Cool and detached from now on. Cool and detached.

14/June

Every day in the house seemed to be a new experience in food. My time in the apartment felt like one long culinary extravaganza. By now I had tried the assorted delights of *galbi*; grilled ribs, *bulgogi*; beef and *samgyeopsal*; pork slices, on the barbeque, which were all wonderful. I also really liked *jajangmyeon,* noodles in a thick black soy and oyster sauce.

Ever since the *ssamjang*, I had realised with some delight that I had stumbled upon an undiscovered jewel in world cuisine. Korean food really was unique. Similar ingredients to Japanese food, without the austerity, and Chinese food, without the preponderance of oils. Plenty of spice, as with South Asian food, but without the curry or coconut. The best way I could think of describing it was like an Asian-Mexican fusion; meaty, with raw flavours and raw spice. I felt spoiled by the generosity of my flatmates, and tried to reciprocate as best I could. I had made some porridge for them at breakfast time over the weekend. It didn't seem like the fairest of exchanges.

Sitting round the table at dinner time, fresh from our studies, eating excellent food, was becoming the staple of the house. That and Ki Duk constantly singing *The O.C.* theme tune:

"*Californiaaaaaaaa*"

The show was the sort of thing I hated; a saccharine setup of beautiful people. But as I started to join in with my friends watching it, I actually started to get into it. I had to admit it was really sharp and well written, with deliciously dramatically taboo teasing plotlines. The girls, coming from a conservative and traditional culture, took it in with shocked disbelief.

"Is American society really like this?"

"No, Ji Eun, this is as crazy for me as it is for you!"

Terrorism

"Conquest breeds hatred, for the conquered live in sorrow. Let us be neither con-
queror or conquered, and live in peace and joy."

— *Siddhartha Gautama*

Since September 2001, "terrorism" has become one of the defining concepts of our times. From everyday discussion to political debate, hardly
a day goes by without this word entering our minds. But what does this
most emotive label actually mean? What are the causes and conditions that
lead to such abhorrent violence? What are the roots of this illness?

A book which aims to present terrorism in the context of our globalised world
for what it is, rather than what it is variously proclaimed to be, is Jamal R.
Nassar's *Globalization & Terrorism: The Migration of Dreams and Nightmares*.
From the introduction, which begins with the author's birth in the Jerusalem
of 1946, a time when the city was in the midst of a terror campaign in which
the names of the terrorist groups were *Irgun* and *Stern* rather than *al-Qaeda*
and *Hamas*, it is clear that this will be a different perspective on the issue than
is commonly available.

Globalisation

Nassar begins by examining the phenomenon of globalisation from a variety
of perspectives; from that of corporate managers and labour unions, governments and investors, environmentalists and protestors. From proponents to
opponents, multiple opinions are presented. Globalisation is therefore understood as a contentious and highly complex topic, but the defining trend of the
movement can be identified as **interdependence**.

Contemporary globalisation is found to be an extension of the trend of
interdependence that has been ongoing throughout history, with the same root
drivers; power, wealth and the pursuit of hegemony.

Following the collapse of the Soviet Union, a single global hegemonic
power has emerged; the United States of America. This geopolitical primacy

has in turn stimulated an increasing homogenisation of world culture according to a singular economic ideology. Nassar discusses the impact of this economic hegemonisation and homogenisation and its contribution to the growing wealth divide between and within nations.

Migration of Dreams

To explore the links between globalisation and violence, Nassar introduces the process of a "migration of dreams" as a direct product of cultural and technological globalisation:

"Take, for example, a remote village in Chad. People there still live a traditional way of life...Imagine if one day the village elder returns from the city with a battery operated television set along with a satellite dish...Watching in amazement, the villagers may see an American, French or Italian movie or some show produced in California, New York, London or Paris...A couple of hours later, the village elder turns the set off, and villagers depart to their mud or clay homes to sleep on their dirt floors. As they sleep, they are likely to have new kinds of dreams. No more are they dreaming about milking the cows or harvesting the crops. Instead, they are likely to dream of having an amazing faucet that brings water inside a home or of having a car or electricity. Once this happens, it sets in a process that one can call a "migration of dreams"...The migration of dreams is a common occurrence. It takes place all the time and in all places."[1]

This change of dreams would lead the villagers to have raised expectations, and Nassar references the social science theory of relative deprivation — which identifies the gap between expectations and achievement as a significant contributor to violence — to illustrate how the globalised dissemination of information may increase the risk of violence throughout the poorer regions of the world.

The rapidity of social, economic and technological change which goes hand-in-hand with globalisation is also identified as a contributor to violence. Such change tends to polarise societies, and polarisation can in turn develop into armed resistance and civil war.

Migration of Nightmares

Terrorism is approached by viewing the concept from a variety of angles; from dictionary definition to common political usage. Nassar concurs with the view that "terrorism" is a politically useful tool and not an adequate research concept.

- The terrorist is always 'The Other', or else a different label, such as freedom fighter, is applied.

For the purposes of the discussion, terrorism is therefore redefined as a political label to describe acts or planned acts of violence for political objectives. With this renewed definition, the acts of armed forces of nation-states can also be encompassed within the label of "terrorism".

As a direct result of the globalisation of armed conflict he introduces a different process; the migration of nightmares. The example of history identifies the terrorism of states and empires as the driving force behind this phenomenon; terror inflicted by the powerful on the weak leads to nightmares of violence amongst the oppressed. The desperate will then in turn inflicted nightmares of violence upon their oppressor:

"Take, for example, the case of Hanadi Jaradat, who carried out a suicide bombing at Maxim's restaurant in Israel, killing herself and twenty-one innocent victims. Hanadi was a well educated 29 year old attorney from the West Bank town of Jenin. Two weeks prior to her deadly attack, Israeli forces came to her family home looking for her brother. When located, her brother was shot and killed. Her fiancé was also killed in the attack. Both were killed while Hanadi watched. It was at that point that Hanadi approached Islamic Jihad activists to volunteer for the suicide bombing. Her subsequent action may have been motivated as much by revenge as by frustration. But her act led to the migration of Palestinian nightmares into Israeli society."[2]

Successes in violent resistance will then in turn encourage others and military responses from the powerful will generate more violence leading to cycles of nightmares. He uses these themes of dreams and nightmares to explore in detail terrorism across the globe in recent times. The situations in Northern

Ireland, Congo, Chechnya and Colombia are reviewed and found to be illustrative of the pattern of domination leading to dispossession, deprivation and violent reaction.

Islamist terrorism is summarised in its full historical and political context. Indeed, the conflict between Israel and Palestine can be seen as the most pressing example of these central themes; the dispossession of the Palestinians and the mandates of Israeli supremacy have resulted in a dangerously volatile situation where the dreams of one group becomes the nightmares of the other, and the inaccurate concept of "terrorism" only serves to obscure the issue:

"Palestinians, uprooted, denied their basic human rights, rendered refugees for decades, and constantly exposed to state terror are collectively labelled by the desperate acts of a few in their midst. On the other hand, the initiators of the terror of occupation, land confiscation, curfews and other illegal restrictions are collectively perceived as the victims of terror."[3]

Nassar's conclusions are stark; that we cannot afford to succumb to fictions and hysteria when discussing terrorism. It must be seen for what it is; as a symptom and not as a disease. A cure for the economic asymmetries and hypocritical abuses of international justice involved in the genesis of terror must be realised if solutions are to be found. Nassar closes with a dream, and speaks of how an informed citizenry could act to bring about an end to terrorism and the causes of terrorism through nonviolence and cooperation, dialogue and reconciliation. To create a more peaceful world, we all must act as part of a global community.

"For hatred can never put an end to hatred; love alone can. This is an unalterable law."

– Siddhartha Gautama

15/June

I invited Jo up to the apartment to meet my friends and have dinner. This caused a lot of amusement amongst my friends who assumed it was a romantic gesture.

"Speak some Korean to her!"

"Sa lang heh" {*Saranghe*} [I love you]

I said, prompting a big laugh from the group.

We went for a walk around the lake in the park by the apartment and talked about our travel plans. Jo had been looking for work through her hostel job club. She and Bernadette were mulling over an opportunity for live-in service staff at a remote upstate bed and breakfast. She wasn't too impressed with my dedication to finding work:

"You don't want to get a job, do you?"

"No. I'm on holiday."

I told her how lucky I felt to have made such great friends and got such a nice place to stay. I didn't see any reason to move on for a while. My last *I-Ching* reading had been "abundance", and I certainly felt abundant; such wonderful new people in my life; loads of parties and the chance to meet other students; the chance to meet up with other travellers through Jo; a great library to work in; the fantastic park we were walking through; amazing weather all the time. She agreed that I was lucky. She had been bowled over with the friendliness of my Korean sharemates. Jo had previously had an idea about teaching English in Asia later in her trip, and now on meeting my friends she was certain she wanted to try it.

17/June

Jo had moved things on in the work front pretty quickly. She and Bernadette had found a better deal than the highway hotel; they had acquired contracts with an international resort chain in the glamorous Whitsunday Islands off the Great Barrier Reef. Nice. They were flying out today. I was happy for her; a great opportunity in a fantastic location. But I was definitely going to miss her.

18/June

This was crunch time for Korea. It was a big gathering in the front room, swelled by the arrival of two of Ki Duk's friends who were staying over for the weekend; a couple who had been working in the resorts up north, and who were now making their way back down to Sydney before going home. They did not have anything good to say about the conditions they were forced to work under nor their Australian bosses. I immediately thought of Jo and her venture into the tourist trade.

South Korea were out, defeated by Switzerland, and the football fever died suddenly with it. After the success of the last World Cup my friends were shell shocked and their enthusiasm for the sport visibly drained out of them as the game came to an end. I tried to cheer them up with my Scotsman's perspective:

"Hey, at least you got there!"

But it was to no avail; they really were a nationalistic people when it came to sports and football wasn't the same experience for them without the Red Devils.

21/June

I had noticed that there had been more frequent comings-and-goings in the house this week. It seemed that every time Ki Duk and Hwani were in, Ji Eun, Bob and Ga Hee were out. And vice versa.

I didn't think anything of it until Ji Eun and Ga Hee conspiratorially asked me what I thought of Ki Duk's friends sleeping over at the weekend. I said I didn't mind. They said they did mind; it wasn't part of the deal that went with the apartment and it had made them uncomfortable.

I decided to keep my own council, and in the evening Ki Duk and Hwani took me aside to tell me that in the house there was a:

"Little problem."

I personally didn't see what the big deal was; Ki Duk helping his friends out with a place to stay was camaraderie in the spirit of backpacking in my view. There seemed to be real anger on the other side though. I think they thought of it as unannounced, uninvited strangers in their home. Thankfully, I hadn't got caught up in the situation. Andy the Waegukin. It seemed that both cliques thought of me as neutral territory, and I was determined to keep it that way. I had an extra long practice session in the park to keep out of the way whilst a house meeting went on. Hwani briefed me on the fall out later that night. Bob, Ji Eun and Ga Hee had all decided to move out once their pre-paid rent had expired.

23/June

I got an email from Jo to say the people, food and locale at the resort were all amazing. However, she was really beginning to distrust Australian employers. She had been lied to yet again. Rather than working five hours a day she was working eight. And that didn't include the unofficial time she was obliged to spend hobnobbing with the guests, an extra five hours on top of the eight. Not only that, she had ten consecutive days to work before getting a day off rather than the previously agreed six. She also didn't have much of a chance to see Bernadette as they worked in different departments. Not surprisingly, she had decided to quit once the month was up.

24/June

A bunch of young men, away from home in a foreign country for the first time, let loose, we were all extremely sexually frustrated. Watching *The O.C.* didn't help. Neither did staying under the same roof with a face that could launch a thousand ships. Our shared frustrations at least provided an opportunity to bond, especially when it was just me, Hwani and Ki Duk hanging out in the hot tub downstairs.

"I want a girlfriend!"

I said.

"Why don't you make a girlfriend?"

A wonderful Korean turn of phrase. *Make* a girlfriend. How do you *make* a girlfriend? I was kind of touched that they had confidence in me to be capable of such magic. I wished I could just *make* one.

"Ji Eun likes you."

I had to admit, it had been something I had considered; she was extremely cute and pert. But there was just something slightly wrong about the idea. It'd be like seducing your best friend's little sister. A loss of innocence for all the wrong reasons.

"Hmmm, too young. Immature."

I asked them what they thought of Ga Hee.

There was some silence.

Then spake Hwani.

"*Laaaaarge Breasts*", said he, with accompanying universal sign language.

I sighed and sank back in the tub as the recollection accompanying Hwani's observation washed over me.

She had breasts like melons; large, pert and shapely. They would have been outsized on a western girl, but were simply sublime coupled with her delicate figure — she had the kind of slenderness that only Asian women can have; elegant and dainty but yet still feminine and curvaceous. There really was no end to her perfection. If you had not seen her you could not imagine her.

They guessed the direction of my queries.

"No, not Ga Hee. Girl like her don't like western guys."

I had watched all the men in the house go through the dynamic of the male encountering the gravity well of a beautiful young woman; approach, followed either by rebuff or invitation to orbit. Ki Duk had taken to occasionally playfully propositioning her in his cheeky outgoing way, and she hadn't been letting him down gently.

The doe-eyed submissiveness that had so struck me on the first day was just an illusion brought on by a combination of Korean culture and protocol and the novelty of first meetings. Underneath the exquisite exterior the girl was pure steel. Ga Hee really was one of life's Queen Bees. The one unfortunate effect she had on the household though was the loss of my best friend Ji Eun. Once the life and soul of the house, Ji Eun had quickly become a mere satellite of Ga Hee, seemingly awed by her model looks and greater ability to accessorise.

I had managed to reign myself in as best I could, in an attempt to make her feel comfortable. But I still found my eyes making contact with hers whenever I moved into the same room. I still found my gaze lingering on her when she wasn't looking. I couldn't help it. Her hair shone, her skin and eyes *glowed*. She moved so well. Her dress was always stylish. The girl was radiant. Feminine perfection embodied.

25/June

I found my Korean sharemates to be very organised and cleanly. Sunday afternoon was official house cleaning time, and it happened every week without fail. Sunday mornings were for Church, and Ki Duk always attended, even persuading some lukewarm others to join in. I had noticed him reading the Bible in the hostel as I was reading my *Gita*. My opinions of Christianity had changed radically over the past year, to understate considerably, but I found Ki Duk to be an example of a true Christian. He really practiced what he preached.

There was a different atmosphere in the house this Sunday, as Bob was leaving for apartments and friendships new. I was disappointed to see him go; I liked him, he was an extremely bright and quick witted young man. Not only that, he was one of those rare people who were better than me at playstation games.

26/June

As one door closes, another opens. Seungri, our new arrival, was a good looking young chap with a very dapper, almost Italian sense of fashion. From first impressions of looking at him I assumed he'd have an attitude, but as I got to know him I realised why I got on so well with most Asian people; they were so genuinely kind and friendly, without guile or pretence.

27/June

The whole city was buzzing for the Australia-Italy match. Although the atmosphere was a little damper in this little corner of Korea, we watched it in the flat and commiserated as our hosts went out. The guys didn't like Italy at all, a hangover of the previous World Cup and Italian gamesmanship in their match against then co-hosts, South Korea.

28/June

Ga Hee was leaving. I didn't think I'd achieved my goal of making friends with her, but the atmosphere in the house had been relatively calm after that first storm. It was a cordial goodbye, and I wished her well for the rest of her time in Australia.

It must have been a day for leavings; Jo emailed me to tell me she'd just quit the resort. She'd had enough of the exploitation and was leaving to Airlie beach. She was going to try and backpack up the coast for a while before heading on to Darwin, but she was now out of funds and down to her credit card. I wished her luck.

30/June

Friday night. Another girl's night out was on the cards. Alicia was over, as sexy and vivacious as ever. The girls were openly flirtatious as we sat around the table, obviously revving up the gears for a night out on the pull;

"I want a boyfriend!" announced Alicia,

"Korean boyfriend ...or Western boyfriend?" cheeked Hwani.

"Western boyfriend!"

She then set her stare on me, like a laser beam, and purred coyly;

"Why don't you be my boyfriend?"

I hoped I wasn't blushing as the girls burst out giggling together. They soon went to get ready in Ji Eun's room, before leaving to the bars and the clubs, looking fantastic and without another glance or word in my direction. *The pain of withdrawn sexual attention.* I knew she had just been playing with me, but it still hurt to feel the dazzle of her gaze, and then to be dropped like a stone for the potentials of the city night.

The guys didn't seem too bothered. Girls who went clubbing were not their type of girls. It was an opportunity to have another Barbie out in the park, *sans* females. Knowing a girl as exotic and as hot as Alicia was out on the prowl while I stayed at home fairly took the wind out of my sails however.

01/July

As I went about eating my cereal first thing, the guys were busy preparing hot rice in the rice making machine. It was something I could never get over; rice for breakfast. It's not as if it didn't feature in every other meal. I couldn't help but noticing a makeup free morning-after Ji Eun looking exceptionally pleased with herself as she went about rice preparation.

Must have pulled? I thought.

Ji Eun sat down beside me with her bowl of rice, and my suspicions were swiftly confirmed:

"Andy ...I need to ask you some questions."

"Did you get a boyfriend Ji Eun?"

She beamed and blushed;

"A *date*. Yes."

"Do you want some advice?"

"Yes."

Everyone leaned in, in expectation of learning more.

"Have you been on a date in Korea?" I didn't really need to ask this, she was a popular girl.

"Yes."

"It won't be much different than a date in Korea, Ji Eun. We're all human."

Even though a World Cup quarter final was on, no one cared. So I was alone to take in Portugal vs. England. It wouldn't be a World Cup without England going out on penalties ...and they didn't disappoint. Happy days.

02/July

Another beautiful winter's day. And another Barbie in the park. Lim's replacement had arrived; Su Bin. She was older than all of us, and we ended up just calling her by her family title; *noona*. Elder sister. I was relegated to being a *dongseng* for the first time. A pretty little lady, she was also the physically smallest of the group which made for a cute juxtaposition of seniority and stature.

Ji Eun was still feeling nervous about her upcoming date. The paternal side of me got the better of my thinking and I decided to warn her about what could happen;

"Ji Eun ...you need to be careful ...my experience of Australian guys is that they tend to be ...how do you say ...*players* ...do you know what I mean?"

Noona and Ji Eun were both looking at me quite intently.

"We have this in Korea too ...we call *Playboy!*"

"Well, a lot of Australian guys are playboys then."

She thoughtfully internalised what I'd said.

I didn't want to dampen her fun but I was genuinely worried. She was such a lovely and innocent young girl and I didn't want her to get hurt. I knew how callous and exploitative Western men could be towards Asian woman. I knew this because I knew myself.

04/July

The revolving door continued to spin, and a new young man had arrived. Won from Busan. He wasted no time in getting straight in to the balcony smoking thing, and started chaining away from arrival onwards. We slowly began to discover more about our new flat mate. His was a tale of woe; in the previous weeks he had just discovered that his girlfriend had been cheating on him. Not only that, cheating on him for months. And it didn't end there; with his supposed *best friend*.

Heartbroken, Won had shaved his head, applied for a visa, and bought the first flight he could get to Australia. And here he was.

The guys thought he looked like the Korean footballer Cha Du Ri. This didn't help Won's despondency:

"Cha Du Ri *ugly!*"

I didn't think he was a bad looking guy, but it summed up how life looked from Won's perspective; everything was down on him. Even with his halting English I could still get a feel for his personality, especially his completely sardonic sense of humour. He could invest considerable sarcasm in anything, through the barest of vocabulary.

Ecological Interdependence

Books have long been our prime repository of knowledge, but some-times a rare book comes along which not only transmits knowledge but has the ability to shift the entire perspective of your mind. One such book in my case was Clive Ponting's *A Green History of the World*. I encountered this work while I travelled in South East Asia, and as I read a whole new conceptual framework for interpreting and understanding the world unfolded...

The first chapter in *A Green History* is devoted to the story of Easter Island, one which is most illustrative of the central theme of the book; the interdependence of humanity and the environment.

Colonisation
- First Polynesians arrive around 5 C.E.
- Perhaps as few as 20–30 people
- They discover a wooded island more marginal for land cultivation and sea harvest than other Pacific volcanic outcrops
- Chicken, taro and yam are their main foodstuffs

High Point

- Written script and complex ceremonial life develop
- Despite resource disadvantages, one of the most advanced societies in history for its technology level emerges
- Population about 7,000 in 1550 C.E.
- Large stone ceremonial platforms, *ahu*, aligned astronomically and topped with imposing statues, *moai*

Descent

- Deforestation for agriculture, heating, domestic construction and monument construction
- Leading to lack of construction materials for seafaring and fishing
- Soil erosion and declining crop yields
- Warfare and population reduction
- Over half the total number of *moai* incomplete and left derelict near quarry

Contact

- The Dutch Admiral Roggeveen becomes the first European to visit the islands in 1722
- He found a society in a primitive state; about 3000 people living in reed huts or caves in a virtually treeless landscape
- The society was engaged in constant warfare and resorting to cannibalism due to meagre food resources

The importance of the metaphor of Easter Island is such that it is also a major component of Jared Diamond's *Collapse: How Societies Choose to Fail or Succeed.* In this masterwork, Diamond takes us on a grand tour of human society, historic and contemporary, examining the factors behind historic social collapse.

Diamond begins by drawing up a five-point framework of factors involved in social collapse:

- Environmental Damage
  o Properties of people
  o Properties of environment

- Climate Change
  o Natural
  o Man Made

- Hostile Neighbours
    - o Involved in historic collapses

- Friendly Trade Partners
    - o Essential goods
    - o Weakened trade partner (environmental factors)

- The response of a society to its environmental problems
    - o Political, economic, social institutions

Using the comparative method of scientific analysis, that is comparing natural situations with respect to differing input and output variables of interest in order to draw conclusions, he applies the framework to a detailed analysis of societies past; from islands of the south pacific, to the Central American Maya, to the Viking colonies of Greenland, demonstrating how the interplay of these factors contributed to the destruction of those particular cultures.

As an antidote to such tragic histories, Diamond reviews how three cultures in similarly difficult circumstances managed to overcome their problems to prosper; the agriculturalists of Highland New Guinea, the inhabitants of the tiny island of Tikopia in the western pacific and the Tokugawa Shogunate of Japan:

Pacific Tikopia

- Cyclic agriculture including rainforest biomimicry
- Managed fisheries
- Zero population growth

Japan 1603–1867

- Cyclic forestry
- Managed fisheries
- Zero population growth

Modern societies and the challenges facing them are also discussed in depth; the diverging fortunes of two nations sharing the same island, Haiti and the Dominican Republic, provided a definitive illustration of the central themes. A window onto the environmental problems facing developing and developed nations is provided by the summary cases of Australia and China, respectively. The author's direct encounters with environmental degradation and its effects on livelihoods in rural Montana humanises that which can easily become abstract, and serves as a reminder that such pressing issues are ongoing within the very heart of the United States of America.

The most harrowing chapter in the book involves the recent genocide in Rwanda and the human environmental impacts involved in creating the conditions for the explosive violence that took place. The Rwandan genocide is seen as an example of what is termed **Malthus' dilemma**:

The reverend Thomas Malthus observed in his *An Essay on the Principle of Population* (1798) that an unchecked population tends to rise at an exponential rate (i.e. 1, 2, 4, 8, 16), whereas food production tends to rise at a linear rate (i.e. 1, 2, 3, 4, 5).

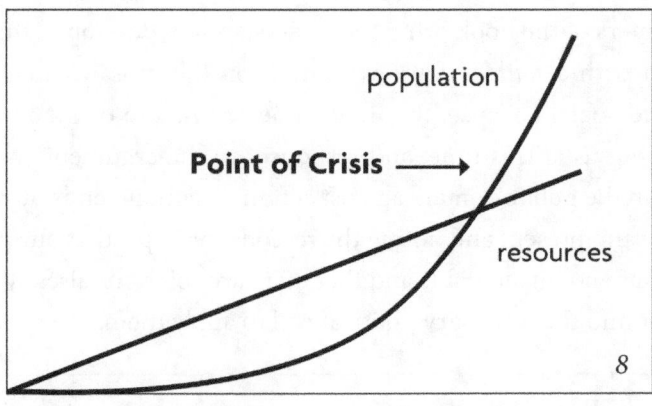

Malthus' Basic Theory

Malthus considered that *preventative* checks, such as birth control, celibacy and late marriage were necessary to prevent a crisis of population outstripping natural resources, lest natural *positive* checks such as war and famine result. A sobering point of reflection, the importance of which is emphasized by the scale of the environmental challenge facing the world today.

The most serious problems are summarised as follows:

1) Accelerating rate of destruction of natural habitats
2) Human dependence on wild food sources
3) Extinction rates of species
4) Rates of soil erosion
5) Limits on fossil fuels approaching
6) Freshwater flow sources almost fully utilised, depleting aquifers
7) Approaching photosynthetic ceiling
8) Global release and build-up of toxic chemicals
9) Effects of alien species transfer
10) Release of atmospheric gases and associated effects
11) Human population growth
12) Rising per-capita impact of human population

The final chapters of the book bring the understanding developed throughout to conclusion with a *tour de force* exposition on the causality behind environmental and social collapse, the interconnected nature of the most pressing issues facing us at this time, and responses to the arguments most often encountered in the public domain against action. Diamond ends by reviewing successes past and present and noting the reasons for hope; that our problems, though great, are not insuperable and that there are solutions already available should we but find the necessary political will to apply them.

Further Reading
Collapse: How Societies Choose to Fail or Succeed by Jared Diamond
A Green History of the World by Clive Ponting

07/July

I'd been emailing Joanna about my lack of work situation, and she had replied telling me to get my butt up the coast. Things were starting to roll for her after the false start at the resort. She was getting a lift to Cape Tribulation courtesy of some Germans she had met, then planned to travel onwards with some Korean contacts she had made. She seemed to be enjoying herself and let me know there was fruit picking on the go in that part of the world...

Back here in Brisbane, things were much the same as usual, though Won was clearly not doing well. His mind was scattered and he often seemed preoccupied and distracted. When he was with us he was mostly chain smoking on the veranda, and he spent a lot of time on his own; burying himself in comics that he kept on a little electronic gadget and music on his headphones. We all felt for him though. His life had just been shattered.

I thought about how beneficial the natural self-clearing of reflection in solitude and the refreshment of walks in nature would be for him. But us moderns tended to take the wrong approach of blanketing ourselves in stimuli in an attempt to forget about the pain.

08/July

We had seaweed soup with our barbeque, out in the park and under the sun. It was Hwani's birthday and the soup was a Korean cultural ritual. It was given to young mothers after childbirth, in order to fortify them after the exertion. In a cute tradition, people drank it celebratory on their birthdays too. It was delicious.

Won was curious about my *qigong* routine, and wondered why I did it;

"We have things like this in Korea, it's called *Brain Breathing*."

Propelling the syllables with a sense of incredulity mixed with disgust. He continued:

"But I think: Stupid. You can't breathe with your brain. *Oxygen!* You know!? *Lungs!*"

I wasn't offended. I knew I never used to believe in such things and I could appreciate his point of view. I didn't see any point in preaching, so I just told him:

"You have to have an open mind, and then try it out for yourself."

Strange that the people who need these things the most are those who are open to it the least.

09/July

Ji Eun was leaving for a new apartment. I'd been surprised she had stayed so much longer than Bob and Ga Hee. Things seemed to be progressing with the new boyfriend and she occasionally asked me for advice; girls asking me for romantic advice a strange turnaround in my world I assure you.

"He keeps asking me: 'Are you happy?'"

"I tend to do that sometimes when I'm out with a girl, Ji Eun; it just means he's a little nervous and wants to make sure you are enjoying yourself."

I was sad to see her go. And I was sad to reflect on how things had turned out; those first weeks sitting round getting to know each other and sharing food felt like being part of a blossoming family. There really was no point in moping though; I'd seen plenty of great friends fall out under the stress of co-habiting, so for it to happen to a temporary collision of strangers was not really surprising. Ji Eun's life, like her relationship, was moving on and I wished her well.

My plans were also rolling into motion. I emailed the documents giving the go ahead for the agency to commence the sale of my apartment back home.

The World Cup final was on, late at night here in Oz. No one cared. Ki Duk and Hwani had gotten jobs working nights cleaning up in bars and clubs and everyone else was in bed. So by myself I watched the artistry of Zidane come to a violent close, and the culmination in a series of magnificent rearguard actions bring home the trophy for Italia.

10/July

With the decision to move on made I got back in contact with the Yoga centre I had contacted when I first arrived in Brisbane. WWOOFing and Yoga sounded like a great opportunity and I wanted to experience it before I left the city. They replied promptly and a placement next week for five days was arranged.

11/July

Studying the sustainability literature always leads to periods of unreality, and I was having one now. The sustainable population estimate of Australia was variously estimated at between 8,000,000[1] and 500,000[2]. The current population was 20 million[3].

I looked out the window at the cars passing by on the motorway on the other side of the river. Life goes on. Oblivious.

Only the other day on the news I had watched a home office minister imploring people to:

"Have more babies!"

The Australia Business Council and the main political parties were aiming for a population of 50 million people by 2100[4].

12/July

I wanted a girlfriend. It had been too long since I'd had someone to hold and someone to share life with. I sometimes thought of Alicia; she was still on the grapevine, still single, and I could probably arrange to bump into her on a night out or at a party. But then I had to admit to myself that I was in no condition to have a girlfriend. I just wasn't happy enough. I had nothing to offer and nothing to give.

Hey baby, wanna come with me to the library?
To hang out in an apartment and watch The O.C.?
Play playstation?

That's pretty much what my life consisted of. I couldn't take a girl anywhere, or show her anything. I couldn't even afford to go out for a meal or a drink.

13/July

Ki Duk and Hwani's truancy was infectious, and Seungri and Won occasionally slacked off from school as well. The flat had a different rhythm now that Ki Duk and Hwani were working nights, and late afternoons I usually found them all clustered out on the narrow balcony, enjoying the sun and puffing away.

We were a contented bunch of escapists. A group of young men, away from home in a foreign country for the first time, let loose; we all just wanted to enjoy ourselves. For me, it was an escape from the relentless grind of nine to five and the demands of life as an employee in a multinational corporation, for them, the crushing pressures of social and parental expectation that comes with being born in South Korea.

High tech on the outside, it was an extremely conservative and traditional culture on the inside, and everything in Korea — ranging from rice to rivers, mankind to mountains — was arranged as a hierarchy. Men were at the top of this paternalistic pyramid, and the eldest sons in each family were expected to shoulder the burdens of the previous generation. Hwani, as an oldest son, felt it the worst of all the guys. But all of them, male and female, felt the heat to some degree; I had heard about the crazy expectations placed upon the shoulders of Japanese children before, and in Korea it was much the same; long hours of school, long hours of post-school study, long hours of weekend extra-school study. Their nerves must have been shredded. No wonder they all smoked. No wonder they all slacked off at the first opportunity.

But we were all currently enjoying this meander in the flow, and life certainly wasn't bad under the sun in Brisbane. The only thing getting us down was sexual frustration. No money, no honey. No one could afford to go out.

I was always intrigued, observing my friends. The differences to me, the similarities. The alien and the familiar.

I sometimes wondered on the other hand what they thought of me; this strange foreigner with his religious *qigong* practice routine? Who preferred their company to that of his own kind? I felt a little like an outcast when I reflected on that point. But then I thought, would I rather be hanging out with a bunch of locals and ex-pats, or having this mind-opening encounter with a new culture and people? There was no contest; I could speak English at home; I could experience Western culture at home. I hadn't reached Korea during my grand tour of East Asia, that revelation of an experience, and there was a certain poetry to finally meeting up with it here.

14/July

One person leaves, one person enters. Ji Eun's departure had freed up another space and Gayoon was our new arrival. A different look and personality; more shy and reserved, but lovely all the same.

I'd informed Ki Duk that I'd be spending the next week WWOOFing, and that I was going to be looking for harvest work soon. My replacement arrived before I left though! Ki Duk's good friend from his time farming, Dong, whom I'd previously met on a few occasions. I found him an impassive little chap; his English was good but his wavelength was different. We had three sharing the bedroom now, but there was plenty of space; the more the merrier.

Dong was big on beer, and the guys indulged big time at his welcoming dinner before settling in to a marathon game of cards. Alcohol often fuels explorations into the mysteries of life, and tonight it was time to question the phenomena of why many Asian women like Western men ...but comparatively few Western women like Asian men. I don't think my friends thought of this as a fair exchange.

"Western women tend to like guys who are big and strong and confident", I told them.

I thought I'd stir their perceptions up a little bit;

"A lot of Western women like Black guys actually."

"*Black guys*?" came the response, tinged with disbelief.

"Yeah; a lot of Black guys are big and outgoing, whereas Asian men tend to be small and shy."

But then I let them in on another secret;

"Actually, a lot of Western girls travelling in Asia told me they found some Asian men handsome and desirable, especially once they met them in person and got to know them. But the problem with most Asian guys is that they are so shy they never make a move ...and consequently nothing happens."

15/July

We were pushing the boat out today. A lot of the old guard would be moving on soon and Ki Duk wanted to do something fun before it was time to go. He and Dong had both bought ancient cars during their time on the farms, and we rolled them out for our excursion down the coast.

We took in one of the main shopping malls on the way, where the girls went wild, before parking up between the skyscrapers on the beach. It wasn't a day for making picture postcards, the wind had whipped the sea into a mist like spray and it was positively cold, but we made the most of it. The girls got dunked in the water and then we had a spontaneous athletics competition, in which I got absolutely smoked by Ki Duk. It was a fun day out, and I loved the way my Korean friends could be so exuberant without any need for alcohol.

16/July

The boys arranged a game of football through mutual friends in a suburban park. An unexpectedly multinational gathering, it was nice to stretch the legs again, and to feel that ease of recovery that *qigong* engendered. As with the athletics, the guys were all much better than me!

Yoga

Day 1

I made my way to the ashram, anticipation and nervousness playing on my mind. Free yoga sounded wonderful, but the strictness of the regime was somewhat intimidating. I was surprised to discover a modern building in an industrial area of town. I was greeted by two young women when I arrived, Shanta and Lakshmi, who were more European than the names suggest. I was shown around the main yoga hall, which was roomy, well aired and light, and the office and living quarters. The place had a noticeable atmosphere; scrupulously clean and well organized, with incense subtly scenting the well-ventilated air.

First order of the day was lunch and we sat in the small garden to eat. Yogic systems were my first introduction to energy and meditation so I was enthusiastic about what the experience might bring, and I was delighted to find that I could have conversations about 'energy' — which become perfectly natural once you learn to cultivate and understand your own — with my new acquaintances. Shanta was keen to promote the ashram and told me all about the setup. Their guru was a teacher from India who had dedicated his life to spreading yoga around the world through setting up not-for-profit centres such as the one I was visiting. *Yoga in Daily Life* was both the name and the ethos of the school, as they sought to make spiritual cultivation relevant to ordinary people and their day to day lives.

I was enthusiastic to learn, and to talk about my own experiences in eastern cultivation. Shanta explained that yoga developed as an aid to meditation. I related that this was also the basis of *Shaolin Qigong* and Kung Fu. The Indian Buddhist Patriarch *Bodhidharma* (founder of the *Ch'an* or *Zen* school) who taught at the *Shaolin* temple in central China transmitted the energy exercises of Sinew Metamorphosis and the Eighteen Lohan Hands to the monks, as when he arrived he found them too weak to practice meditation adequately. These systems over time evolved into *Shaolin Qigong* (Chi Kung) and *Shaolinquan* (Kung Fu) respectively. It is a fact not generally known or understood in the west than spiritual development can only proceed from excellent physical, mental and emotional health.

My first ever yoga session. I settled down on a mat on the floor amongst the arriving students. My *qigong* training was a benefit in getting into the optimal state of mind and body and I could feel little rivers of *qi*, or *prana* as it is known in yogic terminology, starting to flow as I relaxed fully. We went through a routine of relaxation, followed by warm up stretching, then breath cultivation *pranayama* followed by sitting postures, standing postures and then meditation. It was challenging in parts and I really noticed my lack of flexibility. All in all it was an extremely enjoyable and relaxing experience and a discipline which I thought would be nicely complimentary to *qigong* practice.

It was a day of firsts. In the evening I was offered the chance to make my first ever vegan meal for dinner. A fairly straightforward curry, it didn't turn out too badly to my tastes, and the others seemed to enjoy.

Day 2

Up bright and early for meditation during the silent hours of 12 a.m.–8a.m. It wasn't necessary for me to take part but I thought I'd join in with the spirit of the place. I wasn't quite sure how to go about communicating under such a code of conduct, so I tried not to offend my hosts by limiting myself to gestures, nods and smiles.

Morning meditation took place in the main hall and was just myself, Shanta and the other primary member of staff, Pravin. Seated in lotus we spent half an hour so in silence. Lotus meditation is actually a very demanding exercise but I was surprised with how clear and how long I could hold my mind after just a few months of *Shaolin* training. In my school, standing meditation is preferable at the beginners' level with lotus meditation being an advanced exercise due to the potential for harm if performed incorrectly.

Next was 'karma yoga'. When I first heard about it, I was like; *Gee extra Yoga!* But it in fact means: Work. Just work. So I got to work cleaning and mopping the floors...

Lunchtime was pleasant and a chance to meet some of the many volunteers who were involved with the ashram setup. In the afternoon, it was back to karma yoga again. My task this time was a leafleting run around the suburbs. I was becoming something of an expert at posting things through letterboxes. Once again not the most creative or intellectually stimulating of work, but I decided to look on the bright side; a chance to discover parts of the city I'd never otherwise see, and to have a long walk under the bright sunshine and blue skies of Brisbane.

In the evening, another yoga class, this time at the beginner level, and indeed it turned out to be a lot more suitable for a novice like me. Afterwards I had dinner with Brian, another temporary resident of the ashram, originally from Ireland. We laughed at the Brisbane "winter"; 20 degrees and blue sky everyday simply does not fit the definition. He's an interesting conversationalist, keen on philosophy and obviously enjoying life in Australia and his yoga. He confessed that he would probably be spending his days as an alcoholic if he was still living back home. He also had an amusing reflection on meditation:

"By definition the most boring thing in the world."

Which is perhaps true in the beginning, but no one would do it if it were not the tremendous benefits it reaps.

Metaphysics I: Karma

Karma

Metaphysics. Beyond physics; literally the "Big Questions". *Where are we going? Why are we here? What is it all about?* At a less rarefied level; are there processes or forces beyond classical physics involved in shaping human existence? How can we experience 'knowings' beyond the five senses? Why do chance happenings sometimes have such subjective meaning?

An example of a metaphysical model is the yogic concept of karma:

"The Law of Karma says that the energetic vibration arising from each action will one day come back to the one it originated from, wither with the same qualities or even stronger through other intervening interactions".

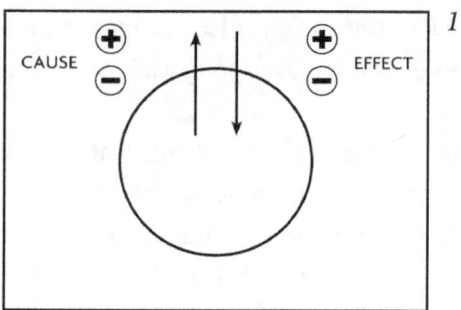

"Karma is a vibration that surrounds our subtle "Phanomen"+. Its fine vibration is not limited spatially and accompanies us everywhere. The effect of our phanomen can be compared with a dynamo which generates two kinds of energy: negative and positive."

The metaphysical model of the yoga schools presents a rather different existential canvass to that of the religious West. In this model, there exist two things; an individual and a Universe. The individual moves through this Universe by the power of thought. Rather than the concept of a Divine Authority which determines reward or punishment based on a prescribed set of laws or judgement, in this model the individual reaps what he/she sows. The basis of morality from this viewpoint then is the understanding of the effects of your own causes. In short, there is only self-responsibility.

+*In yogic metaphysics a term to denote the energy and radiation that surrounds individuals; containing the expression of one's entire spiritual personality.*

Day 3

I visited an internet cafe on my lunch break and received great news. My attendance at my Sifu's (Cantonese for *teacher/father*) eldest daughter's wedding had been confirmed. The decision to attend had come with the decision to sell my apartment back home. I figured if I was going to push the boat out I should at least do it properly; and a trip to Malaysia in October for a celebration and some more high level *qigong* was pushing the boat out in style. The conservative part of me wavered a little at the outlay. But then I thought: *How often do I get invited to a Malaysian Chinese wedding? How often am I in this half of the globe?* I figured I should make the most of the opportunity.

Day 4

The days fell into a similar pattern. The karma yoga kept me busy but wasn't overly laborious or too time consuming. Sitting in the office preparing was leaflets' was good fun, as I got to know the other volunteers as we worked and over lunch afterwards. I also got a chance to check out Pravin's massage studio. Offering treatment along the lines of traditional Indian Medicine, *Ayurveda*, I was delighted to notice he had a steady stream of custom, and to witness how effusive his patients were about the benefits of the sessions. Few things make me happier than to see knowledge of alternative healthcare spreading...

In the evening Shanta, the only full time live-in member of staff, was out visiting friends so I had a quiet night to myself, occupied by the quite wonderful library they had downstairs. A cultural and metaphysical treasure trove, I was in my element.

Metaphysics I: Union

The ultimate aim of yoga is union of the individual consciousness with the divine consciousness, an experience known as Self-Realisation or God-Realisation.

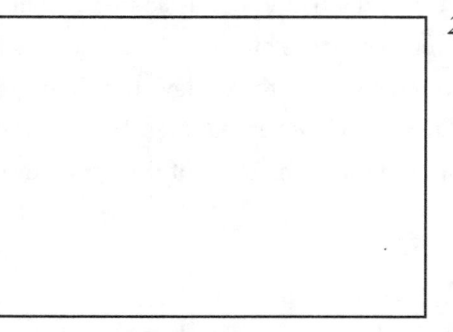

2

"The path that leads us to the experience and realization of this oneness is also called 'Yoga'. In this sense Yoga is described as a scientific system of physical, breathing, concentration and meditation practices, as well as ethical rules for living."

Further Reading
Chakras and Kundalini The Hidden Power in Humans by Swami Maheshwarananda

Day 5

Today was a chance to take part in the advanced yoga class. This class was a little more like my prejudices were expecting; a collection of healthy and good looking people with exceptional flexibility. I hung in there as best as I could, but couldn't quite manage the headstand finale.

I reflected on the wonderful benefits I had received from the week. The metaphysical concept of *vibration* that underlies yogic practice and the ashram way of life had really stimulated my thinking. I had noticed my meditation and the clarity of my mind had improved markedly over the short period of time I was here and I could only put it down to the diet, yoga practice and atmosphere being as optimal as the philosophy underlying it suggested.

I had been incredibly stimulated by all the philosophy and literature I had been reading in the library and I was more than a little curious when invited to join the *satsung* in the evening. I wasn't sure what *satsung* entailed, and I was hopefully half expecting some sort of philosophical exposition or discussion. Disappointingly, the gathering turned out to be *bhakti*; the devotional worship found in temples, churches, synagogues and mosques everywhere. It was not all bad though, later there was a beautiful guided meditation by audiocassette from *Swamiji*, the founder of the *Yoga in Daily Life* movement, which was both relaxing and deeply thought provoking.

Afterwards, one of the yoga teachers asked what I thought of the *bhakti* and whether I had seen anything like it before. I didn't reply because I could not think of a response that would not cause offence. What did he want me to say about what amounted to a sing-song and social club? It was not to my taste, to say the least. But then I mused; perhaps I only hold such an opinion of *bhakti* because I prefer the path of *jnana*?

Day 6:

I said my goodbyes to Shanta and Pravin after meditation and breakfast. Despite such a brief acquaintance I was struck on leaving by how lovely and kind-hearted those two people were.

Wonderful people; the global experience.

As I walked back to the apartment my thoughts returned to "the guys"; my friends from Korea. I realized that I had really missed them over the past few days. The yoga had been a great experience but I definitely couldn't live for a long time in an ashram. A retreat once in a while would be good, but such a restricted life is not for me.

22/July

I got a really nice welcome when I returned to the flat, including a big hug from Hwani. It's only him and Seungri who are about — the rest are away on a day trip to Byron Bay.

I'd gotten a message on my phone from Neil, the instructor in my school I had tried to get in contact with the previous week. I call him back and it turns out he is sitting in the cafe outside the apartment building, which he had just happened to have chosen out of all the cafes in the entire city, after one of his lessons had been cancelled.

We meet up and have a good long chat, and afterwards he offers to take me through a *qigong* session. Mothers with toddlers pass by in bemusement as I roll about on the ground. He advises me to continue as I have been doing.

We say goodbye; it was nice to have met albeit briefly, and I appreciated the opportunity to grill him on future kung fu training.

24/July

I made plans using the maligned *Harvest Hotline*, to do some fruit picking in Bundaberg, a town further up the East Coast.

There had been some household intrigues in my short absence, and not just the arrival of season three of *The O.C.* on DVD. During one of our lad's chats, Ki Duk confessed to his unrequited passion for Gayoon. I had noticed they had been disappearing off together, to go jogging, over the past weeks but I hadn't quite put two and two together. Only problem for Ki Duk was that she had a boyfriend back in Korea. They seemed like they would make a nice couple and I hoped the situation worked out for him as best it could.

30/July

We had had another last supper the night before, pushing the boat out to get some Scotch Whisky. All change: Marissa had died (in *The O.C.*), Noona and Dong were leaving for harvest work and Gayoon was returning to Korea.

Now it was second goodbyes; this time for the last time. I really hope I will be able to see "the guys" again someday. It's a warm farewell.

A leaving lunch for Dong clashes with my bus departure time, but Won *dongseng* graciously walks me to the bus depot and sends me off.

Bundaberg

Seven hours on a bus as a child used to be forever. As an adult in Australia it's just up the road.

I arrived in darkness in Bundaberg. Despite the late hour the bus station is teeming with backpackers. There are a few hostels in view from the station itself, and they look full to the rafters, judging by the crowds of people and the lived in air suggested by the profuse washing lines hanging from the balconies. I immediately began to worry if there will be work available. I located a public phone and called the contact number the *Harvest Hotline* had given me. No response. Nice. I was without a map in a dark open plan Queensland town.

I eventually find a Chinese takeaway and the staff graciously give me directions to the hostel. Pass a well thumbed book exchange shop, the type run by little old ladies in small towns everywhere, and get a thought to have a look in during my time here. I find the hostel, a two story building in corrugated iron that is little more impressive than a tin shack. There are a large crowd of young people, mostly young Asians, milling around in the front patio. I track down Belinda at the front of the throng, the receptionist I had spoken to the previous week and who hadn't answered earlier in the evening. She was being harassed by the crowd, evidently a pay or work dispute, and had the kind of hang dog expression that suggested harassment wasn't a new experience for her.

Eventually I get to speak with her and she confirms my earlier sinking feeling by bluntly informing me that there is no work at the moment. Disappointment doesn't quite cover it, but I figure getting irate isn't going to change the situation. It's late, I've been travelling and I just want to get to bed. I'll leave worrying about the situation to the morning.

My only piece of fortune is that there is one bed still free in the hostel. I say hello to my Japanese roommates; two guys who are busy chatting with one of their friends, a striking Japanese girl wearing bright blue contact lenses. They've been working on the farms this week but inform me it's;

"shit money"

"hard work"

and consequently a;

"shit job"

I'm distinctly unimpressed with the circumstances; the attitude of the staff, the cleanliness of the hostel, the slightly downtrodden air of the tenants. It fulfils pretty much my worst expectations of the itinerant farm work scene. Harvest slavery indeed.

31/July

I wake to find the Japanese guys have returned from an early shift. They listen to my irritation about the situation the *Harvest Hotline* has put me in. My cause for irritation pales in comparison to the fact that they have spent several hours under the tropical sun harvesting zucchinis at $2 per bag. Not little grocery bags either; industrial size containers that take an age to fill and are difficult to lift and carry.

I started to get a picture of my two roommates as I commiserated with them; Tetsuo a short stocky fellow with surprisingly fluent English and a self-confident manner which didn't quite fit with his job as a primary school teacher back home. He was taking a year out to learn English. Yasutaka was more the backpacker stereotype; a student on a gap year with a shy manner and halting English.

I found Belinda again, and tried to make it clear how much of an inconvenience I'd been put to. She was impassive and unapologetic, saying that there had been a mistake with the admin and that she'd give me work whenever it was available. To break my mounting frustration and with nothing else to do, I took a walk around Bundaberg to get my bearings and to do some shopping. The place had the sprawling open plan grid of streets that typifies modern new world towns, large or small. North of the hostel was one of the main arteries of the town, a wide dual carriage main road, an endless strip of signs that disappeared into the heat mirage distance. The side streets leading off the dual carriageway were arranged as a rigid grid of blocks comprising single detached homes. The grid occasionally broke into empty plots, such as on the other side of the railway from the hostel, evidently due to zoning restrictions for commercial, industrial or private purposes currently unrealised. The cells in an accountant's spreadsheet sprung to life. The town layout was completely non-conducive to walking because of the distances involved and the heat of the day; every house an isolated fortress with relationships dependent on car use. I always find these sort of places depressing. The End of Community.

Back at the hostel I got a sense of what was going on. The clientele was a 60/40 split between Asians and Westerners who mostly kept to their own cliques. The work was intermittent, with currently less available than the amount of people staying there. The harvest work itself varied depending on the vegetable of choice, but seemed to be pretty grim; hard physical labour under a baking sun. Giving the whole situation a surreal twist was the recompense for said hard physical labour under a baking sun; it was possible to do a full morning's work (the sun was too hot in the afternoon for work in the fields) and not make enough to pay for the day's food and board!

01/August

Another unconstructive day.

Phoned Hotline.

"There's nothing we can do. We'll get back to you."

Phoned hotline later.

"We'll get back to you."

Asked hostel.

"No work for a few days, we'll let you know."

I did the rounds of all the hostels in the area, and everywhere the same refrain:

"No work this week."

It seemed like I was trapped in this strange little twilight zone of a town. I figured I may as well make the most of it; all that free time was more time to study for my book. I visited the little old ladies bookshop I had passed on the day I arrived — seemed like fate — and found it staffed by a little old lady. I picked up a book called *Seth Speaks*. I sold the *Bhagavad Gita* and bought *Global Warming: The Greenhouse Report*. It was a bit out of date and pretty dry looking but I figured I had to get as good a handle on the issue as possible. The age and content of the book was an irony in itself. I also discovered another side of Bundaberg, parklands and riverside which were great places to relax under the warm sun.

03/August

I was getting to know my fellow hosteller's better as well as the town. There was a big friendly Korean crowd and I enjoyed trying out my newly acquired pidgin Korean on them. In the evening Yasutaka and Tetsuo invite me out for a chance to experience the nightlife of Bundaberg and a game of pool.

The centre of town wasn't far from the hostel and the two story city blocks of the centre loomed out suddenly from the single story monoburb around.

Despite cosmetic differences, the atmosphere reminded me of home. It was much like Thursday night in a depressed area of Scotland — a quiet street populated by occasional drunks. A bunch of teenage girls in a car drove by the intoxicated passengers screaming at any unfortunate bystanders. Strange way to spend your evenings, I thought. Drive around drunk in circles and scream at everyone you pass.

We ended up in the type of bar that is populated by old men all over the world. Dark, quiet and with the usual cast of depreciating alcoholics propping up the counter. We claim a pool table after waiting in the queue but this causes an upset with some new arrivals to the bar; a pair of very large skinheads with a female companion in tow.

We've waited our turn but they seem to be stung by the slight of some Asians getting to play pool before them and kick up a fuss. We ask the staff and other drinkers to clarify the waiting list and they back us up. In response, the bigger of the two guys looms threateningly over Tetsuo, eyes bulging;

"call me a dick then."

"what?"

"If it's my mistake then I must be a dick, so call me a dick."

Tetsuo replies:

"you are a dick"

Extremely bravely calling his bluff.

"Good. That's the way we speak to each other here."

Adrenaline surging at the turn of events, I thought it was all about to kick off. Tetsuo had somehow managed to neutralize the situation with his ballsy reply. I wasn't surprised at this kind of situation developing, and could have imagined it happening back home. In Japan however, I simply couldn't see the reverse and the inner ambassador was stung. Australia must scare the shit out of all the Asians that come here.

Despite this episode, the evening turns out to be really fun. There is an old school jukebox full of rock classics, plenty of beer is consumed and we even push the boat out for a McDonalds on the way home. We exchange stress about our lack-of-women situations in life. Eight months for Tetsuo, seven months for me, two weeks for Yasutaka. Yasutaka had just been dumped by his Korean girlfriend (she went to Surfers Paradise) so that cheered the other two of us up.

05/August

At the weekend a large group of us get a run to the beach from the hostel staff, and I grab the chance to see the Coral Sea for the first time.

The beach is nothing more than a slim spit on the back of small scrubby dunes, not quite as impressive as the expanse and rolling surf of Surfers'. We don't do much more than sunbathe the day away. It's strange that you can be in the tropics, under a clear blue sky, on a beach with some new friends ...but still feel completely flat. We had brought the hostel atmosphere with us. No one was there because they wanted to be there, everyone was there only for the money, and of that money there was not enough to go around. One of the guys had brought a rod and spent the day fishing. He didn't get a single bite.

06/August

Strangest *déjà vu* ever.

07/August

I tried to spend as much time as possible outside the hostel as there wasn't a lot to do beside hanging about the TV room or hanging about the kitchen, which also had a big screen. The Australian version of Big Brother a staple choice and it was the most asinine version of the show I had ever seen, if you can picture that, with a preening cast of plastic people.

I reflected that breaking my TV habit was easily one of the best outcomes of my decision to travel. I was free of the habitual time wasting and the poison of the news. Only when you get outside your own culture and then look back into it do you see how much of a bias is involved in mass mediating events. The news takes the form of a constant ever shifting stream of terrible events that never pauses long enough to look adequately at the causality behind those events. A stream of events filtered both consciously by the manipulation of the broadcaster and unconsciously by the beliefs and assumptions of both the broadcaster and the viewer. Longer term trends such as geopolitical machinations or environmental degradation do not come in for adequate scrutiny. What we get is a constant stream of disconnected symptoms rather than an understanding of disease.

08/August

As the week drew in I was faced with a choice — stay until the fabled 'next week' and its promise, or head on towards Darwin where I had the opportunity to cross paths with some of my old backpacking companions. I figured I could spend the next three or four weeks in Bundaberg working every day with the distinct possibility of not even breaking even, so the decision to cut my losses and move on wasn't hard.

09/August

Belinda showed her compassionate side by not charging me board for the week, which under the circumstances was the least she could do. I was thankful for the gesture anyways and I was happy to be leaving Bundaberg and to be on the road again. Frustrating and unsatisfying in many ways, but like everything there was a sunny side; the people I met, the experience of somewhere completely different and new, the sunshine itself.

Townsville

G lad to be moving on, my mood was sky high as I boarded the bus. The joy of somewhere completely new. I drank in the shifting scenes as the bus moved on; the productive horticultural scenery around Bundaberg gradually made way to thicker forest and lusher greens as we moved north along the Queensland coast, before becoming increasingly dry and scrubby. By the time we reached Airlie beach the coastal landscape had a distinct Grecian tint; bare hills and hot blue sky looking into deep azure Ocean.

Without an income in Brisbane, I had to sacrifice most of the east coast from my journey, with only a stop in Townsville scheduled as I couldn't miss the opportunity the Great Barrier Reef represented. Such expediency also led to the decision to check in to another one of those Aussie innovations — the hostel that is part of the bus station. Handily, it was directly across from the Yongala Dive Shop and I bounced over as soon as I was settled in, eager to book a trip to dive the famous local wreck.

The guy in the shop was friendly but responded wryly to my diving inquiry. It turns out I'm the first customer he's seen in days, and he hadn't been able to take anyone out to the wreck for six months. Obviously bored out of his mind and delighted with the opportunity to speak to someone who was not himself he talks to me at length about the situation. For the entire half-year period, the weather and water flow patterns had been disturbed, generating conditions contra to diving in the locale. He advised me I could take a boat out further into the reef, although this would mean a trip lasting a few days, or instead head further north towards Cairns.

I wandered out of the shop, mind reeling at the implications. Aside from the personal let down of missing out on the Great Barrier Reef, it was yet another example of climatic instability, something that had been presented to me on every stage of my round-the-world trip. If even such minor ripples on planet earth can impact local livelihoods, without even beginning to take into consideration the effect on underwater life — what could the major ripples of our current trajectory do?

10/August

I took a walk round Townsville, a small town in the shadow of a looming edifice named by the colonists as Castle Rock. Another cipher of this continents' divergent history; in any other it would be fortified, here it was just an outcrop of bare stone. With scuba out of the picture I took a ferry to Magnetic Island, the local scenic spot. A pleasant islet of sandy beaches that would recall the southern Mediterranean, were it not for the presence of eucalypts and koalas amongst the scrub. Set in tropical seas it was yet another stunning example of the beauty of Australasia.

In the evenings, the atmosphere of the hostel was as flat as you would expect from a couple of rooms grafted on to a bus station, but I did at least find a fellow Scotsman to talk to. He was a tradesman and had found his skills in considerably more demand than mine, with a corresponding ease of generating income. He was making his way down the coast from Cairns and was really appreciating the difference in climate and way of life between Northern Queensland and Galashiels.

Townsville ➡ Darwin

11/August

The big bus journey. I enjoyed the rising current of anticipation for a new phase of travel as I stowed my luggage, the thrill of discovering the unknown. We head out of Townsville and the rocky terrain and scrubland of the coast gives way to wide undulating plains inland, also covered with low brushy vegetation. The scenery gradually changes from scrub, to plains of dry grassland, through to a wooded savannah that recalls the Africa of lions and wildebeest. I see few animals here though, only the odd cattle. I leaf through the book I bought in Bundaberg....

Global Warming

"We are certain ...that emissions resulting from human activities are substan-
tially increasing the atmospheric concentrations of the greenhouse gases...
These increases will enhance the greenhouse effect, resulting in an average
additional warming of the Earth's surface"

– IPCC scientists (1990)[1]

Global warming burst onto public consciousness in May 1990 with
the unveiling of the United Nations' International Panel on Climate
Change (IPCC) first report. An in-depth assessment of the threat posed
by climate change compiled by 300 of the world's top climate scientists, the
report documented the findings of multiple computer simulations of future
world climate: models which predict that average global surface temperature will increase approximately 1°C by 2030 and 3°C before the end of the
21st century unless greenhouse gas emissions are steeply curbed.

An issue of such unparalleled magnitude deserves the most complete analysis and to this end *Global Warming: The Greenpeace Report* was published.
Commissioned by the eponymous organisation as a shadow document to the
IPCC report, *The Greenpeace Report* reiterated the science of the IPCC working group and cast a critical eye over the weaknesses of the policy responses
suggested by the official body.

2

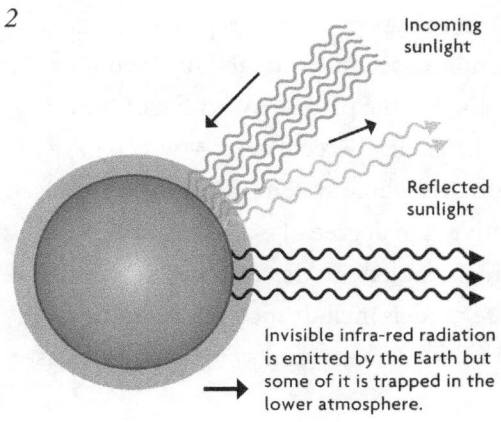

Incoming
sunlight

Reflected
sunlight

Invisible infra-red radiation
is emitted by the Earth but
some of it is trapped in the
lower atmosphere.

How the greenhouse effect
works. Energy from incoming
solar radiation reaches the
Earth. Some is reflected.
Most readily penetrates the
atmosphere and warms the
Earth's surface. Invisible infra-
red radiation is emitted by
the Earth and cools it down.
But some of this infra-red is
trapped by greenhouse gases in
the atmosphere, which acts as
a blanket, keeping the heat in.

Science

The basic science of the greenhouse effect has been known for over a century; solar radiation heats the earth, which then emits that energy back into space in the form of infra-red radiation. Gases which trap infra-red radiation cause some of this heat to be retained in the earth's atmosphere rather than emitted. This natural greenhouse effect is an integral component of the earth's climate system; without it the earth would be inimical to life as we know it, and when at equilibrium absorbed solar radiation is balanced by radiation emitted to space. However, since the industrial revolution, human activities have been greatly increasing the concentration of greenhouse gases in the atmosphere — most significantly the addition of CO_2 through the burning of fossil fuels and deforestation. This increased concentration of gases has led to disequilibrium of the climate system and an <u>enhanced</u> greenhouse effect commonly termed 'global warming'.

Predictions

"It appears likely that, as climate warms, the feedbacks will lead to an overall increase, rather than decrease in natural greenhouse gas abundances. For this reason, climate change is likely to be greater than the estimates we have given."

– IPCC scientists[3]

This statement is underscored by the contributors to *The Greenpeace Report* as one which should have been made much more prominent in the IPCC findings, entailing as it does the probability of positive feedbacks enhancing the scale and rate of warming beyond even the upper estimates given.

Climate results from highly complex systemic interactions involving the atmosphere, oceans, geology and biology of the planet. Warming of the atmosphere and ocean can trigger physical changes — feedbacks — which have the potential to cascade effects throughout the whole system. These feedbacks can dampen (negative) or increase (positive) the degree of warming.

Not all of these feedbacks are included in the existing simulations.[4] Moreover, uncertainties in the climate models include incomplete understanding of the interactions involving natural sources and sinks of greenhouse gases, clouds and polar ice sheets.[5]

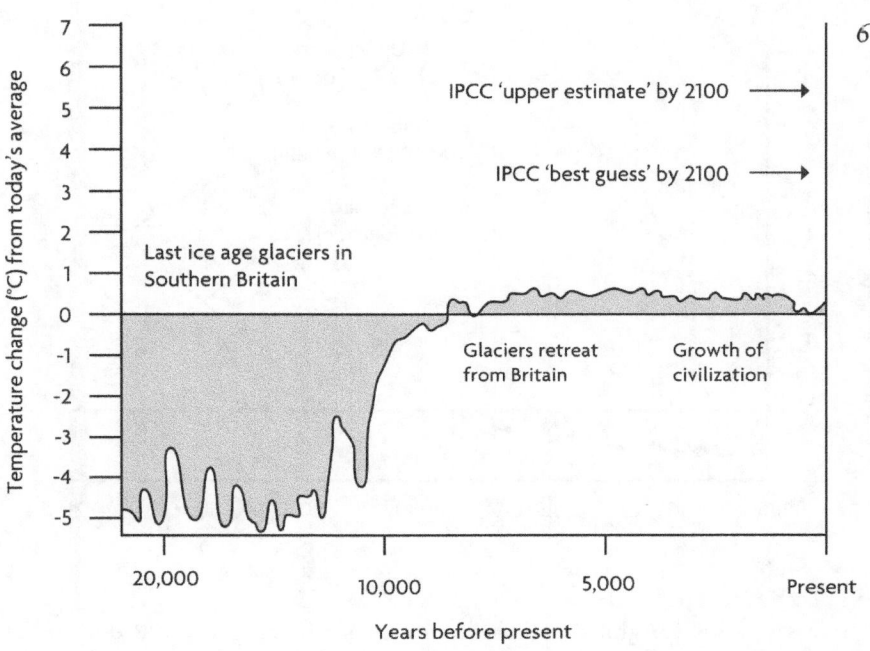

The graphic above demonstrates the unprecedented scale of the estimated changes; to find a similar change in average global temperatures it is necessary to go back to the end of the last ice age. The transition from ice age to our current epoch literally reshaped the face of the planet; involving the retreat of massive glaciation and ice cap cover, a corresponding 100 metre rise in sea levels, migrations of plant species across thousands of kilometres, extreme changes to habitats, species extinctions and species evolution. These major global environmental changes occurred naturally over a time period of 5000 to 10000 years. In comparison, the future changes predicted by the simulations occur over a 100 year period; a rate which would be lost in the width of the right hand margin of the above figure. It is therefore without exaggeration or hyperbole that *The Greenpeace Report* describes the scenario represented by the upper estimates as "extreme catastrophic". It is further sobering to recognise that the upper ranges denoted in the models do not include estimations of temperatures tracks incorporating the positive feedbacks considered by the IPCC as "likely".

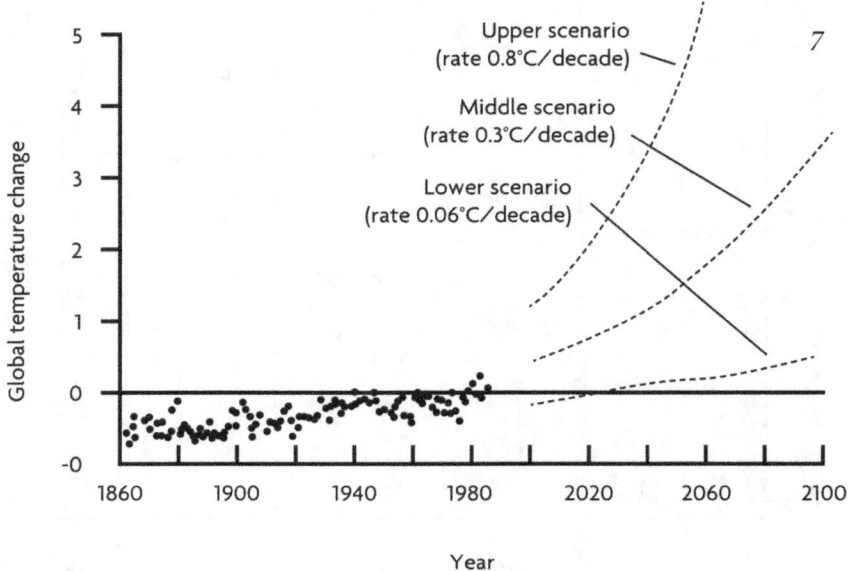

Three scenarios for global temperature change to the year 2100 derived from combining uncertainties in future trace-greenhouse-gas projections with uncertainties of modelling the climate response to those projections. Sustained global temperature changes beyond 2°C (3.6°F) would be unprecedented during the era of human civilization. The middle-to-upper range represent climatic change at a pace ten to one-hundred times faster than typical long-term natural average rates of global change. (Source: *J. Jaeger, Developing Policies for Responding to Climate Change: a summary of the discussions and recommendations of the workshops held in Villach, 28 September to 2 October 1987.*)

The Greenpeace Report summarises that there are many uncertainties, indeed unknowns, involved in predicting future climate; however, the vast weight of scientific evidence suggests that the IPCC findings are accurate. This fact, together with the likelihood of positive feedbacks enhancing warming beyond the given estimates, demonstrates the importance of adopting the 'precautionary principle' in response to the threat of climate change.

Stabilisation Scenarios

"the long lived greenhouse gases (like carbon dioxide) would require immediate reductions in emissions from human activities over 60% to stabilise their concentrations at today's levels".

– IPCC[8]

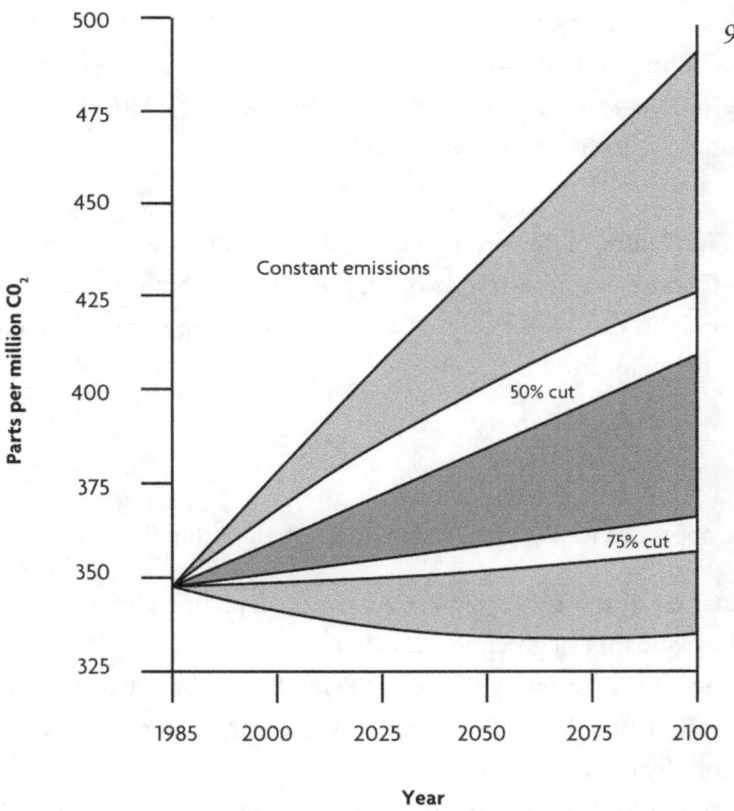

Impact on atmospheric CO_2 concentrations of cutting global carbon dioxide emissions by 50% and 75%, compared to continuing with constant emissions *(at 1985 levels of 5.9 billion tons of carbon per year)*. The spread of values is because two different models of oceanic CO_2 uptake have been assumed.

50 per cent

Proposals to curb global warming are reviewed; the first example explored being a "50 per cent" scenario based on the recommendations of the 1988 Toronto Conference. This consists of:

- Elimination of the production of all ozone depleting chemicals by 1995 and the avoidance of greenhouse substitutes.
- A halt to deforestation by the year 2000.
- The reduction of energy-related emissions of carbon to 80 per cent of 1990 values by the year 2005, to 50 per cent by the year 2030, then constant emissions at 50 per cent beyond 2030.[10]

Computer modelling of these responses predict an equilibrium temperature of 2.6 +/-1.3⁰C above pre-industrial to be reached by 2050, but with temperatures rising beyond this date. It is clear that a more far reaching approach is required to stabilise temperatures.

Global Warming Halted

A possible approach to halt global warming is then outlined:

- Elimination of the production of all ozone depleting chemicals by 1995 and the avoidance of greenhouse substitutes.
- A halt to deforestation by the year 2000, followed by extensive reforestation to offset 1.65 Pg (million metric tons) a year of energy-related carbon emissions by the year 2020.
- A reduction in carbon emissions from fossil-fuel combustion to 30 per cent of the present value by 2020.
- A reduction in the annual rise in methane and nitrous oxide concentrations to 25 per cent of the present value by 2020.[11]

This scenario results in an effective cease to the rise in atmospheric carbon concentration at 445 ppmv by 2030 (only 8 per cent higher than 1990 levels) and a global temperature stabilisation of 2⁰C +/-1⁰C above pre-industrial levels; "goal achieved" in the world of simulations. Of course, the importance of the exercise is in highlighting the radical nature of the responses required to stabilise climate and the urgency of action that will be required.

Policy Responses
Policy responses are addressed in the latter half of the book, with papers detailing the enormous potential of solar power and other forms of renewable energy, the energy savings to be gained from increases in efficiency, and the lack of feasibility of a large scale nuclear fission response to the challenge. This discussion on policy responses concludes with the caveat that unless the underlying driver of human population growth is understood and tackled then progress in averting warming is unlikely; as population increases so does demand for food (thereby increasing deforestation) and energy. Therefore the most significant contribution to reduce the pace of global warming and avoid the worst potential outcomes is the reduction of birth-rates.

Economics
The penultimate chapter is a coruscating critique of our existing economic system and its relation to the ecological crisis by Susan George. Beginning from the root definition of the term economics — from *oikos*, house, and *neimein*, to manage — George discusses grand failings in the economic management of our global house. The role of the lending practices of commercial banks in precipitating the international debt crisis of the 1980s is highlighted, a crisis which brought to attention the activities of the Bretton Woods institutions — the World Bank and the IMF. United Nations Agencies in name, controlled by the richest nations in practice, the effects of their mandatory structural adjustment procedures on indebted nations are identified as the key component in the increasing ecological devastation of the developing world: interest repayments necessitate the conversion of environmental capital into currency. George focuses upon the essential illogic of the ideology behind the development establishment; an ideology which promotes (and enforces) the idea that the entire developing world should aim to emulate the economic performance of the Newly Industrial Countries (NIC) — despite the obvious

real world conditional differences between i.e. predominantly rural Thailand and the city-state of Singapore. Structural adjustment requires export and the result of this policy is an influx of raw materials from the developing nations onto the market, which depresses prices, which leads to further export, and so on; a spiral of resource depletion. The same ideology and practices are also noted to be involved in many of the "aid" projects of the OECD nations.

The "science" of modern economics — which obscures the difference between accumulation of money and maintenance of wealth, which misconstrues efficiency on paper for waste in the real world, and which promotes short-term gain for the minority ahead of benefit for all — is identified as the root issue which must be addressed. George ends by calling for the people of the developed world to set the example and demand that the World Bank and IMF become publicly accountable; for only then will we be able to keep our global house in order.

Editor Jeremy Leggett concludes:
"The uniquely frustrating thing about global warming — to those many people who now see the dangers — is that the solutions are obvious. But there is no denying that enacting them will require paradigm shifts in human behaviour — particularly in the field of cooperation between nation states — which have literally no precedents in human history. That is the challenge for the 1990s. There is no single issue in contemporary human affairs that is of greater importance."[12]

"The Challenge for the 1990s"

I daydreamed of the alternate world that had took the warning seriously; of the massive potential of renewables utilised, of global trading systems made fair and transparent, of organic agriculture enhancing the land, of regenerated forest ecosystems.

A full sixteen years on.

What does the world look like?

What does Australia look like?

Then back to reality. Transport trends towards individuals with gas guzzling SUVs. The trend, globally. The booting up of fossil fuel use across hyperpopulous Asia on a scale that defies imagination. The trend throughout the developing world. Deforestation ongoing, with much more planned. The trend throughout the developing world.

Sixteen years on. What action has been taken? No action has been taken. It is getting far worse.

I finish my daydream and make eye contact with the otherworldly deer of this realm. *Gangurru.* Kanga Roo. I'm delighted with this first sighting proper. They stare blankly back at me, briefly, before the motion of transport carries our gaze apart. The terrain begins to rise from perfect flatness as broken rust coloured hills rear up in the near distance; the shattered remnants of dead ancient mountains, projecting an immense sense of antiquity.

We pull into Mt Isa a full twelve hours after departure for a break stop and switch of drivers. The new driver is a fat loudmouth who tries to inflict what he must consider some sort of stand-up comedy routine via the loudspeaker system. The handful of passengers populating the bus collectively ignore him. Undeterred he continues entertaining himself, honking the horn and pressing up on a car in front that is driving slowly.

"Look, it's a woman! A woman is driving! What did I tell you!"

A wannabee shock-jock, he continues in his nasal tone;

"I make no apologies ladies and gentlemen; I'm a misogynist bastard!"

The bus takes a detour as some of the roads are closed due to the Mt Isa Mardi Gras, an oxymoronic image if ever there was one, and we head out along un-surfaced roads, our displacement of clouds of red dust illuminated by the lights of the bus. *Cocktail* is the film of the night; as a barometer of the 1980's it was not pretty.

I drift off into a fitful bus seat sleep as the driver guns it across the middle of nowhere.

12/August

03:45 a.m. Tennant Creek
A change of buses once more. I shiver in the night air as I wait between the loading and unloading. It is amazing with how quiet the setup is. On the map there is one big line bisecting the country but rather than the continental artery you would expect the remoteness of the route is illustrated by the unkempt, approaching obsolescence buses, mostly full of unoccupied seats, and the single rest stops after hundreds of kilometres. There is only one other couple who are foreigners like me, the rest mostly Aboriginal people. It's nice to see the natives of the land, if surprising because there are so few of them elsewhere in Australia.

Doze off uncomfortably before first light wakes me. The scenery is a beautiful surprise; similar to the wooded savannah across inland Queensland but more thickly forested and interspersed with red termite mounds. At the breakfast stop enjoy a walk around to stretch my legs. This part of the Northern Territory is not as much of a desert as my prejudices expected or the view from the air anticipated; there is moisture in the morning air and life all around. There are even occasional pools of water alongside the road.

The wooded savannah continues as we progress towards Darwin, the forest gradually becoming slightly thicker and lusher, the red termite mounds that are scattered through the savannah at one interval replaced by larger brown ones. The terrain begins to roll until about 50 kilometres from the coast when it opens out onto wide plains punctuated with immense broad leaved trees. The view from the bus then becomes increasingly industrialized and urban.

Two young Aboriginal women had been on the bus since Townsville and were now sitting across the aisle from me. I had spent a lot of time thinking about talking to them. But then I worried, would they see my attempt to break the ice as friendly or condescending? Would it appear as a sexually motivated approach? What would the other passengers on the bus think?

They left the bus before Darwin and I mentally kicked myself, frustrated at my own stupid inhibitions. Afterwards, I could only think about how much joy, fun, friendship and learning I had experienced in Asia, in similar situations where people had talked to me just because they saw me as different and were curious.

As I reflected on the behaviour between ethnicities I had seen throughout the journey, there did appear to be a strange mutually enforced apartheid going on between the Europeans and the Aboriginals. This made sense the more I observed it and when I reflected on descriptions of social contact in Aboriginal culture. Rather than applying a simple label like "racist", it was more to do with social signalling; the body language was subtly different, with the most obvious gap being the fact that the Aboriginal people did not make eye contact according to the same protocol as Europeans and consequently appeared shy and withdrawn to me and the others.

Darwin

No room at the Inn. Thanks to a star crossed collusion of what appeared to be all of the major sports played in Australia happening simultaneously in Darwin, every single hostel and hotel in the town was fully booked. It was the first time in seventeen months of travelling that had happened to me.

Fortunately, before sleeping rough became an option I managed to contact my friends in town. They were generous enough to smuggle me and my stuff into their dorm. They paired up in a single bed allowing me somewhere to crash for the night. I was grateful, and the gesture summed up the type of people they were. Peter and Gavriel were an English couple and our paths had first crossed in Beijing. We had gotten to know each other along the way of the backpacker trail in China and I ended up meeting with them again in Vietnam, Thailand (twice) and now here in Australia.

Kinda redefines the term stalking, doesn't it?

They were wonderful people and I was overjoyed to see them again.

We head to the local backpacker establishment (we got vouchers with the hostel rooms!) and have a great time catching up over some beers. They are in the middle of their Northern Territory tour and are effervescent over the experience they've just had. Lucky my friends are, and I realised they had got *that guide*. THE Guide. The one I had heard about from Relinde way back in Sydney; the one who had stayed in Arnhem Land and had been initiated into many of the secrets of the Aboriginal people. During Peter and Gavriel's tour, he had shone a rich light onto the culture, the history and the life of the natives.

My friends have also noticed the strange Aboriginal-European thing going on, quoth Peter:

"They both ignore each other and no one ever mentions the fact that they both ignore each other."

13/August

The rendezvous was brief as Peter and Gavriel were heading out at dawn for a tour of *Kakadu*. I phone Jo and arrange to meet up in the evening, at the local festival where she's been working.

I spend the morning looking about Darwin. It's a glorious sunny day and I take to the atmosphere of the place immediately. Those recent days on a bus had really underlined the frontier nature of Australian society, and this place capped it off with its distinct outpost feel. It may be a noticeable dot on world maps, but in the flesh it was a most diminutive city, literally a few city blocks worth of Sydney dropped in the middle of nowhere, exceptionally remote from the bulk of the Australian populace.

In the evening, I took a walk through the Mindle beach market, a small collection of food and assorted bric-a-brac stalls set by the beach that makes up the eastward sea boundary of the promontory Darwin sits on. I grazed happily on the panoply of Eurasian snack delights as I took in the eclectic sights and sounds of the busy little market. With live music in the background it was the perfect place to relax and watch the red sun set in a purple sky.

There were a lot more Aboriginal people in Darwin than everywhere else I had so far visited, so much so that it actually began to look like it might be their own country. I wondered what my fellow tourists in Darwin thought of them. I personally hadn't seen a greater juxtaposition of humanity than between the European settlers and the natives.

I had an eye-opening illustration of difference in mindsets between the two peoples as I first entered Darwin on the bus. Some Aboriginal people were sitting beneath a large spreading tree that was isolated on an island of grass between several road carriageways. It was a sight that struck me because the layout of the roadways was a stereotype of suburban town planning of the type that can be found almost everywhere on earth these days. Almost everywhere, trees surrounded by concrete rivers constitute dead spaces that people never habitually use.

I realized that when my city mind interpreted the space it only saw the roads, whereas the Aboriginal mind must have been attributing more significance to the great spreading tree and the grass. I saw numerous examples of different use of environment like this during my time in Darwin, but the perfect illustration was the sight of two families from the differing ethnicities sitting near one another in the parkland around the market. The Aboriginal women sat on the ground with their baby beside them on a blanket, by contrast the Europeans sat on stools, above and separate from the earth, and their baby was clutching at the bars of the cage in which it was penned.

My overall impression was of an extension of subtle patterns of behaviour I had witnessed across South East Asia, whether from the people of the various hill tribes or rural Vietnamese, Cambodians and Thais. Patterns of waking and sleeping, use of time and space, and relationship to the natural environment indicative of being much more closely tuned to the rhythms of nature than the Western expatriates and tourists like me, who took their industrial timekeeping and worldview with them wherever they went. The only difference being that the Aboriginals were attuned to nature to a greater degree than any other people I had previously encountered.

After nightfall I crossed the road to the Botanic Gardens where the Darwin Festival was ongoing. It was quite a lovely setup; clearings in the park had been decorated with artwork, many of the massive trees bedecked with lights and decoration, giving the main amphitheatre the illusion of floating in an organic sea of evening lights. It was that rare type of post-colonial cultural exhibition that successfully marries dignity, respect, novel creativity and professionalism.

I happened into Jo almost as soon as I walked in to the place and we caught up over dinner at the food stands. I remarked over her change in weight — she had only just recovered from an infection of parasites she had picked up while camping on Moreton Bay. Jo's harvest season had been almost as much of a non-starter as mine, but in cutting her losses she had found the ultimate WWOOF. I was really happy for her; when you join the WWOOF movement you are really looking for "it", whatever that means for you. A unique place, a way of life, a set of friends, a shared experience; *connection*. Jo had found her "it"; staying with a collection of granola families, artists by trade, on a hundred acre farm in the mountains of the sunshine coast hinterland. The experience had been relaxing, creative and fun, and had left her uplifted and energised. It really was what she deserved, and I was also glad to find out that the WWOOF agency had immediately struck off the hosts who had abused her way back in May.

As Jo was on duty I took in the show of the evening by myself; an Indonesian cultural song and dance. It was quite wonderful and the hypnotic *Silat* — a martial art of the Malays — demonstration at the interval gave me further inspiration, if any were needed, to take up *Shaolinquan* proper. There was a jam session following the show and Jo joined me after her shift had finished.

14/August

It was Jo's day off so we had a chance to hang out together. I really enjoy her company; she was the only traveller I had met here in Oz that seemed anywhere close to me philosophically and we have a good day just talking about stuff.

We find a greasy spoon café for dinner and a long conversation about environmental issues ensues. We are both optimists and believe that people will eventually wake up to how serious the situation is. The only question being:

"How bad are things going to get?"

"That's the scary part."

"MMmmm ...but it's a hard thing to criticise, I mean, here we are..." I continued, hinting at our circumstances and pointing around at the passing automobiles and electrical appliances surrounding us;

"...right in the middle of it."

"Yeah. But that's the only way it could be." she responded.

The discussion gets increasingly silly and fantastical, the way hypotheticals can sometimes do, with Jo crowning the tangent with one of her hitherto undisclosed theories:

"Everything's got to be *scientific*. And the ruling scientists should have the power to enforce their laws properly."

They tried that, I thought, it was called Communism. I tried to explain my stance as best I could.

"Mmmm ...I don't really think along those lines. In fact, I think that is getting to the heart of the issue. One group of people trying to control all of the others."

Later, as I lay in bed that night the earlier conversation kept coming back to me. *Scientific*. That word could be employed in such a variety of ways. As a *qigong* practitioner I now knew that there were a lot of perfectly scientific things in the world which would be denied, denigrated and ridiculed by certain highly influential opinion forming groups using the invocation of the word *Science*.

15/August

Peter and Gavriel are back from their tour but ready to move on again, this time a flight to Perth. We only have time for a long morning's breakfast together. Despite our crossing of paths being such a brief interlude here in Darwin, it was more than worth it and I would have driven with that crazy bus driver for two days across the outback all over again for the chance to meet them. A married couple, medic and geologist by training who had known each other since Uni, they were in the middle of a long world trip that had been bisected and part funded by Peter doing hospital work in Melbourne. Peter was as high as a kite after *Kakadu*, trying to tell me as much as he could without spoiling it. I had not really anticipated much about the National Park, but the light in his eyes suggested something truly special. Gavriel was browsing the London papers and occasionally interjecting the best bits into our conversation:

"Member of British National Party complains of public intolerance."

They were wonderful people; so free. Down to earth and fun their company was always a pleasure.

We reflected on the almost inexpressible changes in our outlook on life travel had brought about, in contrast to humdrum work routines.

"*A Career*. That's the biggest lie we were ever sold. *Nobody* has a career."

Gavriel recounted the life of people from home who thought they had a career; working overtime during the week to 'get ahead', constantly stressed, a prisoner to their work. Spending Saturday buying expensive clothes then going out and getting hammered at night, Sunday hangover. Repeat. Living for the weekend, justifying the hangovers and some toys because they "worked so hard" during the week.

Living for the weekend.

The term stirred something deep and long forgotten in me, like a childhood nightmare. I couldn't believe I had ever lived that way. The possibility of return to such existence caused me to shiver. The time came for them to leave we said goodbye warmly. I was sure I'd see them again but doubted it would be on this trip.

In the evening I met up with Jo and took in the festival once more. An Aboriginal group called *Red Flag* are playing. The performance is very impressive, the story of the song and dance recounting the contact between Maccassan sailors from what is now called Indonesia and Aboriginal people many, many moons ago.

Live for the weekend? I'd rather live for every breath.

Outback

"Walking is good.
You follow track......
you sleep,
wake in the morning to birds,
maybe kookaburra.
You feel country."

– Bill Neidjie, Bunitj clan

Day 1

It's not yet daylight as we load the van and trailer for the two day excursion into *Kakadu* National Park. The clientele are a diverse bunch — young tourists from Ireland, Switzerland, Holland and Canada, an Australian housewife, a white girl from Guadeloupe, a Japanese man, an American military veteran — and we go through the getting-to-know-yous. I'm sitting beside a retiree from Sydney who is on an adventurous backpacking journey from Darwin to Perth.

Distances in Australia are on a massive scale and we drive for a few hours before we enter the Mary River Park for our first activity of the day; a boat tour along the eponymous river. We've travelled a long way from Darwin and the wetlands of the river stretch out as far as the eye can see in every direction. It is quite an incredible habitat, reminiscent of the Everglades of Florida, and the panorama encircling the horizon is a rippling disc of vegetation and water. We spend the morning cruising along the river, taking in the quite astonishing concentration of wildlife, most noticeably the crocodilians. Whether the slim-line fish-eating freshwater species or the larger, heavier, everything-eating salt-water version, the place positively teems.

Teeming is an adjective that could be applied to any of the creatures inhabiting the river side, which is in constant motion with life. The iconic Jabberoo, a type of giant stork that is another fine example of the Alice-in-Wonderland nature of Australasian birdlife, are often to be seen wading in the

shallows, regarding us passing tourists with disdain from their frozen ped-
estals. The river is an ornithologists dream; egrets and various other small
wading fowl more fecund than my knowledge of their names allows description patrolled the shallows, pelicans fished, and kites and other majestic rap-
tors wheeled in the air or perched in the scattered trees by the riverside. When
the reeds gave way, the banks alongside the river were often the territory of
sun-bathing crocs. There were even a few brave wallabies to be seen foraging
near the riverside, apparently nonplussed by the nearby engines of predation.
The concentration of wildlife is such that the cruise feels like taking part in a
Discovery Channel production.

We are all left dumbfounded by the spectacle of the Mary River, and the excitement builds as we continued on towards *Kakadu* National Park proper. The
landscape changes, crests of bare rock rise up in the distance, and eventually
we reach the park entrance and begin the long drive across vast floodplains
towards our destination, *Ubirr*, one of the most iconic sites of Aboriginal heritage. Our guide narrates the story of the park and its human and natural
history as we roll along the vast distances of the lowlands. The now familiar termite-mound strewn wooded savannah composes the main body of the
landscape, a savannah which is intercut by the vast floodplains of the river
systems that course from the valleys of the ancient time-eroded plateaus that
border the south and east.

An Introduction to Aboriginal Thought — Part I

Systems at work

Yarralin people tell us that the earth is alive and is constantly giving life, the mother of us all. The fact of one mother makes us all kin of a sort....

In Dreaming ecology there is a political economy of intersubjectivity embedded in a system that has no centre. The essential points are:

- *The system is self-contained and self-regulating*
- *Parts are interconnected*
- *It is not necessary for every part to be in constant communication with every other part because information from each part stimulates actions which are themselves information for other parts*
- *The system has the potential to get out of balance and to be brought back into balance*
- *There is no hierarchy, no central agency*

Everything comes out of the earth by Dreaming; everything knows itself, its place, its relationships to other portions of the cosmos. Every living thing has, and knows, its own Law. The result is a set of interrelated parts which is always in a state of flux. When the cosmos is punyu it is homeostatic...The system works, as a system, because (i) its parts are conscious, because (ii) they communicate, because (iii) they act and react, and because (iv) they adhere, as a matter of self-interest and free will, to the same set of understandings.

The process can be seen in the seasonal cycle. The relationship between sun (often identified with femaleness) and rain (often identified with maleness) can be set out diagrammatically; if A (sun) then B (rain); if B then A. In ordinary time we experience sequence: A->B->A, but in Dreaming the relationship is simply A<—>B. Sun and rain, and by association, female and male, exemplify the fundamental feature of **symmetrical complementarity**. We can see here all four meta-rules which I have suggested underly this concept of system. Sun and rain are autonomous, each having its own Law. They respond to each other's actions. In opposing each other they sustain each other. Their opposition is antagonistic in that while one is abroad the other is eclipsed, yet this is an antagonism in which if one were to 'win' by annihilating the other, all of life would be lost.

Each 'part' is both part of the total system and a system in itself. The most significant point to this concept of systems within systems is that there is no hierarchy: the same meta-rules apply to all.

There is a cultural relativism in these concepts that is pervasive and elegant: all parts of the system have their own worldview. That one's own view may be most important to one's life does not mean that the world is focused on humans as a species or on one country and Dreamings over and above others. An essential part of human culture is to know that other Dreamings and other parts have their own news. Once one understands, one can learn the system from any point. The trick is to know what one is encountering.

Every part is cross-cut by others. Matrilineal identities (*ngurlu*), for example, tie people into different nodes. People who share a *ngurlu* are part of a category which includes that species. Emu people share their flesh with emus. If an emu person dies, other people are reluctant to shoot emus because this group has suffered a loss. In this way a specifically human or country viewpoint is enlarged; the enlargement is never uniform for every identity is cross cut by others.

Our view of the monotony of termite mounds and trees begins to change as our guide informs of the Aboriginal understandings of the landscape; common plants take on the appearance of medicines, diverse food groups, and daily utensils, an understanding of the relationships between plants and animals allows insight into the whereabouts of these creatures, and above all a sense of appreciation for the sustenance of nature emerges.

Ubirr is an outcrop of rock in the centre of the lowlands, a cardinal point in the geography of the region. A site sacred to the tribes of this region for millennia, the rocks of *Ubirr* are decorated in wonderful examples of native artistry, and our guide discusses the fascinating history of their origin and development, their meanings and culture. I'm struck by the mysticism of the painting methodology and I detect a metaphysical meaning underneath the conventional explanations given by the tour guide. I'm absolutely thrilled when my personal re-discovery is validated in the words of the creators themselves:

"abundance magic"[1]

We climb up the rocky path of the pinnacle of *Ubirr*, past many other examples of rock artwork until we come to the spot made famous by the film *Crocodile Dundee*. From a hundred feet up or so, the northern savannah stretches into the horizon towards the out-of-sight sea. Celluloid can't capture the reality though, a truly stunning scene of primeval aesthetic, and I have to admit that previously I could have only imagined such natural majesty in an African context. Further up on the outcrop affords a 360 degree view of the park, and we all gaze about enraptured. The immense floodplains stretch out in front, and behind the rocky Arnhem land plateau rises, forming the horizon border to the South and East. *Mesozoic* is the word I would use to describe it; dinosaurs would not look out of place in such rugged ancient beauty. Whispy columns of smoke rose occasionally in the distance, as park rangers continued the land maintenance begun by the original inhabitants.

Throughout the day I'd been getting to know my fellow travellers little by little. I had spent a bit of time talking to Akinobu, a Japanese man who worked as a photo journalist for Reuters. He was combining a short break with his work. His life story was rather dramatic; he and his family had lived in New York prior to 2002 and his wife had worked on the 79th floor of one of the towers of the World Trade Centre.

On September 11th 2001, she was very fortunately getting her breakfast in the canteen on the ground floor when the plane struck. The majority of her co-workers were not so lucky. Understandably traumatised, the family left New York soon after. They were now based in Singapore, although Akinobu didn't like the place, which he considered;

"Too boring!"

And

"A despotic little Chinese kingdom!"

In true Japanese fashion, he apologised for divulging such fervent and personal opinions so soon after meeting. I told him not to worry, and I again reflected on one of those wonders of travel; of how your own personal history begins to interweave with History itself.

The sun begins to drop towards the horizon but we press on for our final destination before camp; a bathing spot that has been certified as crocodile-free. It is a long path on foot and dusk falls as we make our way across difficult stony terrain. The notice boards with getting-eaten disclaimers add to the humorous banter that flows before we prepare to brave the watery realms of the wild. Once we get to the large quiet pool at the head of the stream, nesting between brownish orange rock, I'm in before I know it, the clear water a siren after a dusty day in a truck full of sweating people.

The water is positively delicious; warm enough not to hurt, cold enough to refresh. Eventually everyone overcomes their crocodile fear and we swim lazily around the pool, up to the small waterfall that is its source. Beautiful pastel shades stretch slowly across the sunset sky, the sun itself long since invisible in this steep sided valley. Stars start to wink into view as dusk settles. Eventually we have to tear ourselves away and begin the drive to the campsite for dinner.

After our evening barbeque I go through my *qigong* practice in a quiet wooded spot behind the campsite. In meditation the purpose is to quieten the mind and let go of your thoughts but as my *qi* flow begins my mind becomes absolutely alight with inspiration which I just let flow. All of the new knowledge developed throughout the day is coming together and I feel that I have been given a glimpse into a completely different view of the world. The super-abundance of life, the Aboriginal perspective of the landscape which becomes a living larder, apothecary and tool shed combined, the astounding natural beauty, the sense of spirituality and purpose interwoven throughout. As I contemplated my breath, the eternal communion between inside and outside, I could only think that for the people who once lived here it must have felt as if they had never left the womb. Existence completely supported. I was awash with a sense of peace and joy as I looked up at the dark sky in the soft warmth of the night air to the light of innumerable distant suns.

Day 2

We begin the day with a gentle cruise down one of the rivers that flow through the park. The river cuts deep into the escarpment at this southern section, forming a canyon which tunnels into the plateau. Faulted ferrous cliffs rise from the water's edge, towering vertically on both sides. Eventually the river becomes too boulder strewn to be navigable and we climb out for a walk along the creek side to *Twin Falls*, the cataract at the head of the valley.

The waterfall is immensely beautiful. At this time of year, the dry season, the flow is relatively weak but this does not by any means reduce the sense of majesty of the place. We sit on the soft sandy beach below the pool and soak up the sight. The head of the valley is a large horseshoe, with the path of the falls falling from the left hand side of the "shoe" rather than the centre, as you would expect. The water pools round in a loop beneath before progressing downstream through the boulder strewn narrow channel behind us. The scale and sense of elemental power in the place is incredible. Around the massively eroded head of the falls the remaining cliffs stand resolute like sentinels, the rock aglow in the sunshine. It's a wondrous spectacle which gives the impression of a vast natural cathedral.

Our second activity of the day is a trip to another of the great falls of *Kakadu* — *Jim Jim*. The route to this set of falls is similar; a long walk along the side of a river too shallow and boulder strewn to be navigable. It's a beautiful walk along clear water amid the cool shade of the eucalypts that line the valley floor. The vertically cliffs begin to enclose as we traverse along. As we come closer to the falls, millennia of eroded rock fills the upper channel of the valley like a profusion of so many marbles. Marbles ranging from an average of six foot or so in diameter, up to the size of houses. Respect has to be given to the park staff for maintaining the naturalism of the scenery but it's extremely treacherous underfoot conditions, especially for the older members of our party. This is underlined as we pass one group of rangers who are carrying an unfortunate leg break casualty out on a stretcher.

We reach the head of the valley, another deep horseshoe and find *Jim Jim* a mere dry season trickle. Once again, this does not affect the spectacle. The cliffs rise in a crescent hundreds of metres vertically into the air, creating a great natural amphitheatre. I am absolutely overawed with its beauty. All of the eclectic collection of tourists around, from my party and various others, are similarly stunned by the magnificence. Religious metaphor in hushed tones drifts on the air as we seek for words to describe.

The horseshoe pool of the valley is huge, deep, and also extremely chilly due to the lack of sunlight caused by the towering height of the enveloping cliffs. There is a lot of fun to be had just trying to get in. I have to sit this one out as I had injured one of my toes in the morning dark of the campsite, and I didn't fancy the combination of tropical water and open wound. I feel rather left out as I watch some of the others swim out to the trickle of the waterfall, which is more like rain after it's downwards journey, then climb the cliff side to jump back into the pool.

I find a spot in the sun, relax on my rock, and enjoy the beauty of creation. The sense of Life in this place is *astounding*, even though it is mostly dead rock and water. The thought; "The Universe as a Product of Consciousness" runs through my head like a mantra. In our modern worldview this is a strange and nonsensical statement to make perhaps, the opposite view; "Consciousness is a Product of the Universe" is surely the common assumption.

But it is the discovery at the heart of all religions; the root of your own being is the same order of thing as the root of All Being. As I looked around, I could not distinguish the trees and life in the valley from the rock and the water. All shaped one another and did not exist independently. It was a palpably spiritual place and again the sense of being in a cathedral, though possessed of grandeur far greater than any man-made edifice.

And the only worship required was the simple act of being alive.

Afterwards we spend some time relaxing on a beach downstream, warming up in the sun away from the shade of the deep cliff walls. Our time in *Kakadu* is drawing to a close; four of us including myself are returning to Darwin for the evening before continuing on with a separate group to Alice Springs. Despite the brief nature of the trip it's been a great experience and we all wish each other well as the two parties go their separate ways.

Last Evening in Darwin

As part of the tour we got free accommodation and a free dinner at the staple backpacker bar of Darwin. The feed wasn't bad and afterwards there were quizzes and other party games onstage, to which Ellen, a Canadian member of our group got quite involved. I left early to try and catch up with Jo at the Darwin Festival. The festival was a short taxi journey away and as we passed a group of Aboriginals the driver points and exclaims gleefully:

"When there's too many we round them up and throw them out of town!"

Then grins at me as if for approval. The guy was in every other respect a normal looking middle aged Australian citizen of European origin and I'm shocked at the casual inhumanity of the statement; discussing fellow human beings as if they were some variety of particularly naughty sheep. I had to reconfigure my opinion of the voluntary mutual apartheid; I still thought that cultural and body language differences were part of the reason, but I had to admit that I simply did not know what the Aboriginals thought of Europeans — it was quite probable that they were resentful of being made strangers in their own land. And if average European colonists like this one thought of their fellow people in sub-human terms then this would constitute much of their side of the barrier.

Meet up with Jo and we have a beer together as she's not on duty. We take in the jam on the main stage at the end of one of the shows. The clientele attending the festivities was a diverse crowd, including many local Aboriginal people. Under the influence of music, dance and shared enjoyment the barrier between the two cultures which had been so obvious all the way from Townsville completely vanished. I had been very impressed by the Festival and the sight of the two peoples dancing together and enjoying one another's company was the perfect farewell to Darwin.

Jo gives me a lift back to the hostel in the Volkswagen camper van that she was sharing with a housemate. If anyone could pull off a Volkswagen camper it was Jo. She seemed to be doing fantastically well in Darwin after a mixed time up the coast and I was happy to see her so happy. It seemed like our Australia travel plans would no longer overlap and it was only as I gave her a hug good-bye that I realised with some pathos that we might never see each other again as long as we both shall live. Travel friendships are bittersweet.

Day 3

Another start at sharp a.m. Painful and bracing in equal measure. It's a bigger truck and bigger group for this section of the tour, as we head south towards the Katherine Gorge. A few familiar faces from the past days, but mostly a new, younger crowd. With a morning schedule consisting of driving in more or less a straight line for hundreds of km, our guide breaks the monotony of desert travelling with a pop and general knowledge quiz. Dividing the teams by sex and setting a prize of lunch-making duties fairly fires the enthusiasm and fierce competition between the boys and girls ensues. There's a noticeably different feel to this group; more like an 18–30 tour than the previous rather genteel arrangement. The quiz is a great way to break the ice and personalities begin to emerge as the rounds go on and the stakes begin to rise.

By the time we pull in to our camp male sex have achieved a resounding victory — assured by a points tally in the music of all rounds — and lunch duties are spared. Afterwards, the bus heads down to the gorge where we have the option of a few different activities, though most of us pick the rafting. I team up in a two man canoe with Ellen from Canada, whom I'd met on the *Kakadu* trip. We have two hours to explore the gorge, and the small flotilla of boats from our group surges competitively upstream. Tiredness kicks in pretty quickly and most of us then take a more leisurely pace upriver. I get to know Ellen better as the day progresses; a young schoolteacher from Canada taking advantage of the holiday time such a job provides. It's another stunning Australian day and splashing about in a river the perfect recreation in the heat.

The gorge is shallow and the river easily accessible at the pier side, but soon steep rocky cliffs rise on each bank. The canyon winds and twists along its length, providing fresh motivation to cross each set of rapids as a new section of the river comes into view. Each spectacle provides a new thrill of discovery as the channel becomes deeper and wilder as we progress. The rapids have to be traversed by carrying the canoes on foot and it takes us two attempts and much furious fighting against the current to cross the second cataract. A bunch of us take some time out from heavy lifting to explore the gorge walls and look for the Aboriginal rock art that still remains in the area. It's a spectacular place; another fine example of the primal beauty of the outback; deep blue sky, rust red rocks, olive vegetation and dark water. Eventually time begins to run out and the third and final cataract will have to remain unconquered. We have some fun on the way back down, "surfing" the falls, and at one point I get flipped into the water; an absolutely delicious experience after the work of carrying canoes in the heat of the day. We time our return journey well and spend the route back paddling lazily or floating with the current under the shade of the trees. A delightful afternoon.

In the evening it was time for games around the campfire, the usual sort of things familiar to scout and guide camps. I noticed our guide was taking some sadistic pleasure in putting people on the spot and humiliating them in public, like an overgrown school bully. I decided to retire early.

Day 4:

A long day of driving across the *Never Never*. Ellen volunteers her music for the bus stereo, a bizarre little racket that can be described as Ottawan garage nu-metal, and which goes down very badly with the rest of the passengers on the bus. For our morning break we stop at a park with some thermal springs and enjoy a dip in a warm river, slightly reeking of rotten eggs.

There is also a swimming spot at the confluence point with the local non-thermal river, bedecked by intrusive "Crocodile Warning" and "Swim at your Own Risk" signs. A standoff with the murky blackness ensues before the most brave/stupid of us decide to jump in. It's refreshing but once in the water, which is completely occluded by the admixing of the two streams, a primal fear emerges and no one can get out again fast enough.

Further down the Stuart Highway we reach Daly Waters pub, a little place which pretty much nails the stereotype of the outback watering hole. A social oasis, the place was bedecked with a raft of memorabilia, testament to all the life that had gone through here. I found it a nice opportunity to walk around and stretch the legs but I wasn't particularly fussed about the place. I spent some time talking to Tim, a Danish journalist from the *Aftonbladet* that was also on the tour. He was like me, a quiet personality not fully involved with the rest of the group. Must be something to do with the writers' mindset I thought; being a detached observer on life. Tim was a great guy once you got him talking; very thoughtful and intelligent.

After many, many more miles we stopped for the evening at a campsite on a cattle station. This evening we were "Swagging". The famous anthem *Waltzing Matilda* is an ode to a swag — a kind of heavy outdoor sleeping back used instead of a tent. As it hardly ever rains in the desert and semi-desert of the Outback, tents are not necessary and the one man swag a much more portable and practical option. What I and everyone else wanted to know was the practicality of such an arrangement in a country famed for its array of toxic and occasionally downright deadly creepy crawlies. We were assured that said creepy crawlies much prefer the dark spaces of tents and outhouses than a zipped up swag out in the open; a major factor in their popularity.

Post evening meal we had an opportunity to do some stargazing, and our guide showed another side of his personality in his enthusiasm for the heavens and his knowledge of the Aboriginal legends of the night sky. In this remote place in the southern hemisphere the stars are simply incredible, something I have never seen before; infinitesimal in number, enriched by constellations majestic, the great galactic sweep of the milky way a river flowing across the sky. It was yet another advantage to the swag; falling asleep with heaven as your mirror.

Day 5:

After rolling onwards across the rusted plains our bus pulls in with several other tour parties for a chance to see the Devil's Marbles. I had been looking forward to this part of the journey since the brochure and the sight didn't disappoint. Outcrops of rouge rock dotted the vast savannah plain, weathered down to numberless singular boulders like immense ferrous eggs. Scattered like — well, marbles — across the landscape, often balancing one upon the other, they provided the perfect photographic opportunity to capture the deep reds of the land and the deep blue of the sky. It was great fun to explore.

We continued on for hours across the vastness of the interior, stopping in at another historic oasis, the Barrow Creek pub. We also had the opportunity to visit an Aboriginal market. More villages and townships had began to appear roadside as we closed in on Alice Springs, and it was shocking to see the Aboriginal people in such a sorry state; they appeared listless and downtrodden, often to be seen wandering in groups with tell tale paper-wrapped bottles in hand.

Eventually ranges of hills reared on either side of the road and we passed through the pass to the Springs. I had been expecting little more than an outpost, but the town was surprisingly sprawling and industrial. We had an evening free to explore and with some relief I took the chance to spend some time on my own. I hadn't been enjoying this phase of the journey as much as *Kakadu* — the personality of our guide grated and I had to admit to myself that I felt tired of the relentless revolving door of people into and out of my life. From the hill in the centre of town I watched the sun set over Alice Springs; a beautiful sight as the last light played over the mountains and the stars began to populate the sky.

An Introduction to Aboriginal Thought — Part II

An Introduction to Aboriginal Thought — Part II

An angle of perception is a boundary, and boundaries are both necessary and arbitrary. Necessity lies in the fact that there are no relationships unless there are parts, and without relationships there is only uniformity or chaos. Arbitrariness lies in the fact that since all parts are ultimately interconnected, the particular boundary drawn at a given point is only one of many possible boundaries.

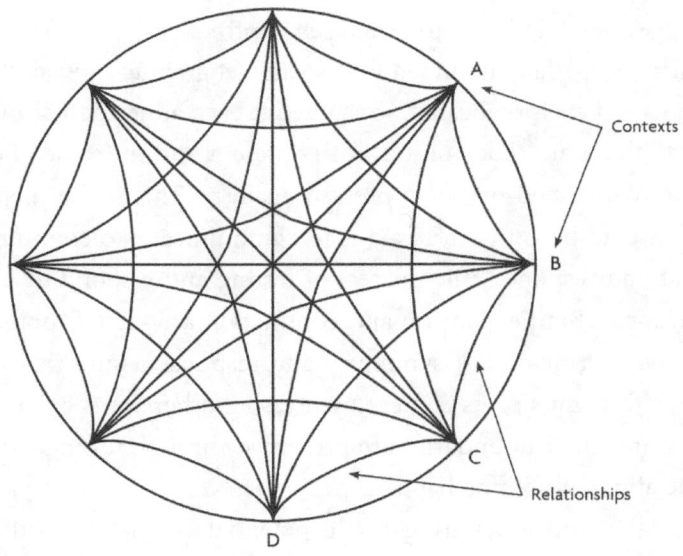

Figure A

Each line in <u>Figure A</u> is both a boundary and a relationship. Each node (A, B, C, etc) is both a context and an angle of vision, another centre. The view of the system changes from centre to centre. To be wise, as I understand it, is to know that there is only viewpoint. Our particular human angle defines our world as it is because it is we who are looking.

Perception distorts, but wisdom lies in knowing that distortion is not understanding.

A country and its people can be bounded in such a way that it can be seen as a closed system. In doing this we see the intimate relationships between people and country: the food of the country grows people, and people return life in protecting the country, enhancing its fertility and giving part of their dead body back to country. The flesh of humans flows into country just as surely as the food that flows from country gives us flesh, and Dreamings, themselves autonomous beings, are there in country. The viewpoint of country is like the viewpoint of any other part: it is unique, self-centred, and interconnected to other like units which are also unique and self-centred.

Earth is spatially bounded into social/geographic units each with its own Law. Law is replicated, so that while each adheres to its own, all adhere to the same. Each unit is **both** different **and** the same. Boundaries are maintained by being pressed against. This is the meta-rule of response: to be is to act; to act is to communicate; to communicate is to test and respond. The process of testing and responding affirms relationships. Both expansion and contraction are seen as potentially destructive. Principles of symmetry and response assure that as any part tests the limits of its context by pressing others, it is balanced by a return pressure. Boundaries are preserved through testing, with the ultimate aim that nothing happen.

That the process of testing has the potential to generate hostilities is a fact of life, just as it is a fact that one way in which parts depend on each other is that they eat each other. The music of the spheres is not a sweet harmony; it is made up of assertive, and potentially hostile, statements of self. When Yarralin people are deeply grieved or deeply affronted, they rise up in anger, saying that only blood will satisfy them. The emotionally satisfying response to one's own loss is to inflict a loss. Balance is not only an abstract meta-rule; it is the content of emotional life expressed in daily exchanges, politics, ceremonies, in life and in death.

The other side of self-assertion is co-operation. Context is never absolute; boundaries cross-cut each other, and can be drawn broadly or narrowly. The boundary between men and women, for example, is also a relationship. It involves the co-operative management of human life, but it is also competitive. Like boundaries between countries, gender differences are cross-cut by ties of kinship and marriage. One of the most significant ties is that between brother and sister. These two share the same body, country and Dreamings; they share the same viewpoint, and a threat to one is a threat to the other. Another highly significant tie is that between husband and wife, who are, by definition, different. Yet their tie is deep, often because of their emotional commitment, but minimally because they share rights to a new generation of people...

As a total system the cosmos accommodates both male and female, and in accommodating both denies the priority or singularity of either. This quality carries through to various beings within the system such as the rainbow Snake. It carries through various social domains such as the cross-cutting of boundaries. It carries through to human beings: all people share in the same burdens and privileges of being human.

Everything in the cosmos is potentially whole. I want to be clear: I am not saying that wholeness is a context that integrates a dualistic cosmos. I am saying that wholeness is a fundamental quality — so fundamental that it characterises the separate parts of the system, including human society and human individuals. To be a person, in the end, is to be simply that — neither singularly male nor female, but wholly human. We are dealing with a most basic assertion that everything is, at the same time, a singularity, a multiplicity, and a whole.

Yarralin people play with boundaries, contexts, and identities; they play with almost any intellectual proposition which can be played with. But they do not challenge fundamental order. Dreaming strings are networks of being and of value; each node is a matrix, and the people and other living things at any given place see themselves as the centre.

For them it is. But they also know, and assent to the fact, that each centre is one among many and that none dominates.

Stanner expressed with great elegance his understanding of the
Aboriginal concern for balance:

"I do not wish to create an impression of a social life without egotism, without vitality, without cross-purposes, or without conflict. Indeed, there is plenty of all, as there is of malice, enmity, bad faith, and
violence...But this essential humanity exists, and runs its course, within
a system whose first principle is the preservation of balance. And, arch-
ing over it all, is the logos of the Dreaming...Equilibrium ennobled is
'abidingness.'"

Excerpt from:
Dingo Makes Us Human: Life and Land in Aboriginal Australia by
Deborah Bird Rose

Day 6:

Another tour group and another feel. A different guide and only a few remaining faces for this phase of the journey, one that I was highly anticipating; King's
Canyon, *Kata Tjuta* and the exalted siren of the centre; *Uluru*. Our small bus
began the deceptively long journey out across empty desert roads and our new
guide, a tall ex-stockman, narrated the changing terrain with a characterful
mix of easy sense of humour and hang dog demeanour.

After several hours driving through varied terrain of shrubby desert,
wooded savannah and impressive mesa outcrops reminiscent of the American
South West we arrived at the Canyon. It was a challenging climb up the side
of the incline in the rising heat of the day, but a rewarding one as a spectacular view of the surrounding plains came into view as we progressed up to the
north roof of the canyon. Passing through formations of weathered rock as
we followed the circuitous route to the top of the cliff walls, we were treated
to the incomparable aesthetic of the red centre. Everything from the rock to

the plant life to the dashing animals to the deep blue sky seemed to be secretly participating in creating a totally unique take of earthly life in this particular place. Silver and black, olive and lime, red and blue, it dazzled the senses with its novelty and its sharpness. It was like exploring another planet.

After traversing through a narrow channel of bright red rock we came out atop the north cliffs, providing a spreading view across the canyon as it twisted west, and the expanse of the plains we had came from to the east. We were high up, and edging out to look down the sheer faces was a reacquaintance with vertigo. Our guide with a sense of national pride explained the difference between a geological canyon and a gorge and pointed out that the States' Grand Canyon was nothing of the sort.

We hiked onwards across the roof, marvelling at the epic erosion of the badlands of the surrounding plateau and then down into the imaginatively christened Garden of Eden, a small water filled valley, before climbing again, up the cliffs of the South Walls. From atop this vantage deeper into the canyon the spectacle of its full length can be seen; a deep fault filled with millennia of scree, the sides topped by vertical towering walls. The immediate sections of the overhanging cliffs are sheer naked rock, salmon pink as opposed to rust red, betraying their origins in recent cataclysmic rock falls. Huge shattered boulders fill the gorge below. There is a palpable sense of geological dynamism in the valley, temporarily frozen-in-tension, which questions the wisdom of dwelling for long on these steep overhanging walls.

Bedazzled and gently exhausted by the hike we lunch and then drive on to our camp for the evening. On the way our driver skids to avoid a tiny little animal on the road. Flushed with excitement, he jumps out to rescue the little fella from its predicament. It's a spiny devil, one of the most beautiful reptiles I've ever seen, rare enough to be a treat even for an experienced outdoorsman like our guide. We all get a chance to get close to the marvellous wee creature and marvel at its dazzling desert camouflage and coat of thorns. It was a great discovery to end a day of great discovery, but we are in for even more as we reach our campsite, which is situated in full view of Uluru. There is still light enough to see and it doesn't disappoint. Even at a distance it dominates the surround — a deep orange monolith in the fading sun, set in the middle of horizon spanning plains. It is different than what I expect; more knotted and varied in shape and contour, banded across with streaks of erosion. Dappled by light and shade it has a curious living quality.

We have some relaxed chat over the campfire after dinner, and are joined by several of the guides from the campsite. The topic shifts to Aboriginal culture and one of the younger guys tells us of the series of initiations that a child goes through on the journey to and through adulthood. He explains that their culture is almost a total opposite of our own; introverted as opposed to extroverted. As an example children are deemed ready to be told secrets of the tribe not when they ask questions, but when they *stop* asking them. To the elders, this is a sign that the child is internally thinking things through rather than expecting all knowledge to come from outside. It was an amazing discussion, and it was equally amazing to hear these, what on the outside, appear your typical outgoing, beer drinking Aussie blokes talking with such genuine reverence and awe about the Aboriginal people and their culture. Off duty and having a smoke, their opinions came from the heart and were not part of the professional tourist guide patter.

Once again it is time to sleep and we have a choice of tent or swag. There's no contest and I bask in the beauty of the stars until unconsciousness takes me.

Day 7:

Uluru. Uluru. Another lifelong dream comes true. We drive to the starting point where we have the option of climbing up to the top or taking the walk around the base. It's an offence to the beliefs of the Aboriginal owners to climb the massif but I'm stunned to find hundreds of tourists from all over the world parading up the path regardless. There is a large signpost at the foot of the trail which is split into two halves; one half which contains the Aboriginal people's request for visitors to walk around rather than climb, the other the safety considerations to be observed when doing the climb. Tourists cluster around reading the safety details, yet simultaneously completely ignore the Aboriginal message. It as if one half of the sign doesn't exist; they don't even read it, straining all the harder at the safety instructions so the other half doesn't come into their field of vision. Then up the path they go. It was a picture of such forced awkwardness that I thought about taking a picture to commemorate it. But almost immediately I relented. I didn't want to immortalise such idiocy. Such a memento would make me want to smash my camera.

The sight drives me furious and I curse vehemently under my breath, more than surprising some of my companions with my anger. You don't go to Rome and deface the Sistine Chapel, you don't go to Mecca and piss on the Ka'ab. But the feelings of the Aboriginals regarding their most sacred sites do not count and do not matter; they are simply ignored.

They are ignored because the Aboriginal people have no power.

Our driver asks who in our tour group are climbing, then with some disbelief notices than no one will be. He brightens up at how "culturally cool" we all are. The tour group splits up as we begin the walk around, everyone setting their own place and drinking in the experience in their own time. Sunrise is on its way and as the darkness turns to light the rock of Uluru begins to unveil its colour in the first spreading rays. There is a real sense of tranquillity in the morning of the life-filled desert, and a hushed awe of expectation as we watch the colours change as dawn draws closer.

The sun comes fully over the horizon and the rock becomes deep orange in reply. It *glows*. It's a breathtaking sight to see the earth mirror the sun. The colour is vibrant and vital, like blood. The organic quality of Uluru comes to mind again; the pock-marked face of the rock, now basking in the morning sun, is reminiscent of the matrix of bone marrow. Up close the rock is nothing like a monolithic lump; richly varied to look at, composed of smoothly textured and banded sediment. Carved by both water and wind, its sides are covered in gashes and rents exposing the honeycombed innards of the rock.

We learn many details of the religious significance of the rock as we walk around. Its caves, gullies and projections are places of great ritual significance in the transmission of 'man's knowledge' and 'woman's knowledge' in the Aboriginal culture; associating intimate sexual geography with the geography of the massif. All throughout the outback tour; from *Kakadu* to Alice, us tourists had been lapping up all we could of the Aboriginal culture. No matter how disparate in character, nationality and background we were, we all had this common point of interest, and no one could get enough. We were all on tenterhooks whenever our guides let us in on the stories, myths and knowledge of the ancient people of this land, and these exchanges had a rarity that made them precious. It was a rarity that seemed very fitting in view of the Aboriginal culture's traditional emphasis on initiation and *secrets*. It truly felt we were being let in on secrets.

I spend most time chatting to Ellen as we make our way around. Ellen, a Catholic, doesn't follow my sentiments on climbing as:

"God is everywhere."

She underlines her point by skipping up and down the sides of Uluru as she goes. For me it's not about realities but about respect. I wondered how she would react to an Aboriginal person violating the sacristy of her local church.

Conflicting opinions aside it was a beautiful morning's walk. *Uluru* changed with the sun, moving from deep orange to a pale ochre against the deep blue sky. Its varied topography holding discoveries around every corner; from rock art to even an unexpected pool of standing water, fed by rainwater channels from the top. It was easy to understand the significance it holds for the local people; it was unique, truly the centrepiece of this great country. We were all quietly awed by the time we finished our circuit and assembled for the next phase of the journey.

Kata Tjuta, *Uluru*'s lesser known sibling was the next destination and one that I was really looking forward to. Ayer's Rock was a postcard staple but *Kata Tjuta* had always been a mystery for me. Another protusion of rock massif through the desert soil, *Kata Tjuta* was equally culturally significant. As we approached we could see that it was also equally visually significant; a series of rocky projections stretching across the desert rather than a singular monolith, not as high but far more expansive in size.

The up-thrusting rock of *Kata Tjuta* creates a series of narrow valleys as the outcrops intersperse with the flat desert land, and we have a chance to explore one of the larger of these valleys — the *Walpa* Gorge — on foot. The sight of the towering hills is reminiscent of its brother, though the sediments of the rock are composed of coarser sands and its smoother sides are a darker reddish hue.

As we draw closer to the valley nestling in between the great towers of rock, a sense of pilgrimage descends on our party. There is a tangible natural quiet, and the sight of the rays of the sun delineated in contrast to the shadow projected by the steep rock sides of the gorge fills us with a hushed awe. Once inside the valley, we cross the shadow that cuts out the desert sun as sharply as a limin. The whole U-shape of the deep defile is surrounded in blackness, cool and deep and in incredible contrast to the burning haze of the open plains we have just came from. As we admire and reflect, the ancient ceremonies of this "Men's place" seem but a whisper away.

Our expedition complete, we head back to Alice. There's a leaving dinner at the hostel but it turns out to be a fairly flat affair; the temporary encounters of the tour not enough to generate lasting friendships in this instance. A few different people from different places, now going in different directions. Once again I enjoy the unregimented time on my own at the end of the day, the last day of my time in the great red centre.

Alice ➡ Melbourne

Another loooong bus journey.

I regarded the passing brushy redness now with a sense of familiarity. As the sun set over the desert I reflected on my time on the outback tour. I could only think: *Magnificent*. The past seven days had been very definitely one of the most magnificent experiences of my life; a series of natural wonders and marvels the equal of anywhere on earth. But there was something nagging at me. Beneath it all, there had been something missing.

I had no one to share it with.

The journey continued on in darkness and unsuccessful attempts to sleep. As the early morning light spread out across the desert flats a strange sight was presented to my eyes. For miles upon miles along the road the surrounding landscape was nothing but a vast flat plain pockmarked with conical slagheaps. A surreal moonscape of reversed craters. Stranger still, there was habitation out here on this blasted wasteland. Temporary structures littered the surrounds, more like tents than buildings, mirroring in shape and form the heaps of the mine tailings. Underground dwellings? It didn't seem real or human. It was like *Tatouinne*.

The bus stopped for a rest break at a station in the 'town' located in the centre of the unnatural desolation. I had a wander about to stretch my legs. *Coober Pedy*; the opal capital of the world. There was an Asian guy, a fellow passenger, also having a walk across the other side of the courtyard. In the manner of travellers on long bus journeys everywhere, we were both taking the opportunity to stretch our legs and have a brief look around somewhere totally new and completely different. We exchanged glances momentarily, our eyes communicating:

What the fuck is this place?

I got a cup of tea and a sandwich, thoroughly weirded out by the thought of people living somewhere like here for any length of time.

24/August

Adelaide arrived in the early morning. I had twelve hours to spend before continuing on to Melbourne. I checked my bag into a safe and went for a wander about. The cold and damp hit me immediately as I stepped out of the station, the sight of leafless deciduous trees slightly later. It was an almost unbelievable contrast. I had arrived in what felt like a European winter a mere bus journey after the baking heat of a desert.

The climate reminded me of home, as did the city of Adelaide itself. The city planning was curious. The central district was comprised of a north-south grid of Victorian-era stone buildings. Surrounding this, as the sea surrounds an island, was a circle of parkland. The parkland itself was also regimentally planned, and housed a wealth of cultural amenities, such as museums and art galleries. I'd never seen anything quite like it, and I took it all in as best I could in my totally zonked state. Sleeping on buses really does not refresh.

The briefest taste of Adelaide, then onwards towards my destination of Melbourne. It was early evening when the bus pulled away and the journey through South Australia and across Victoria passed in almost complete darkness. As I thought forward to Melbourne and Tasmania, I realized I was moving on to the final phase of my time in Australia. I played with the possibilities in my mind. What did I really want out of it? What did I want to happen?

Wouldn't it be nice if...

...I could make some good friends?
...find some work?
...maybe even meet a girlfriend?

Melbourne

25/August

Melbourne immediately impressed. Previously I could never shake the association of *Neighbours* with the city, but as I took my first steps out of the station I was greeted with a beautiful riverside waterfront and the impressive skyline of a world class metropolis. I made my way over to the south side of the river, looking for the hostel I'd booked going by online reviews. When I found it I was heartened; not too far from the city centre, well priced with good clean facilities, and that all important ingredient; free breakfast.

28/August

I had allowed myself a weekend of recovery from travel, followed by a day of orientation. I reacquainted myself with the recruitment agency. In a surreal twist, the girl who had interviewed me in Sydney had re-located to Melbourne in the intervening period, and she interviewed me again.

29/August

I planned to not beat myself up worrying about income for the first few weeks as I had when I had first arrived in Australia. Just relax for the first few days was the plan. I was chilling in my bed, engaged in reverie, considering ways to divert myself whilst settling in, when I was rudely interrupted by Bastian, the new arrival from Germany who had been allocated the bunk underneath mine.

"Hey man, come down!"

"Hey man, let's go get dinner!"

He was a short fellow whose height served to emphasise the cherubic quality of his blue eyed, blonde-framed head. I found his badgering on very first meeting to be pretty pushy to say the least ...but he had an energetic good humour to him and so I decided to give him the benefit of the doubt.

Bastian had literally just got off the plane, and was as high as a kite on adrenaline just as I now realised I had been way back in March. He was the same age as me, and had quit his job to do a working holiday out here. He was a big talker, and was bursting at the seams with ideas for business ventures in the New World. His enthusiasm was catching, and his dream-orientation struck a chord.

He was also obsessed with cars, which wasn't quite me, but each to their own.

30/August

I came back from another morning of fruitless job searching to be greeted by;
"Hey man, I've got a car!"
What?
Bastian had bought an ancient second hand scrap heap for a fistful of dollars. I thought during our first conversation that he was joking when he was talking about getting a car, or that he would at least wait until he'd earned some money first, but no, he'd just went out and bought one. He insisted on taking me for a spin straight away and we went on a big tour of the town in the bold red Bastian-mobile.

Melbourne impressed more and more, from the wide streets of the city centre, to the trendy vibe of the south beach.

31/August

"Hey man, I've got you a job!"
What?
"I've got us both jobs, come downstairs, look!"
On the noticeboard was a sign-up list for two days of warehouse CD packing, to which he'd already added both our names. I didn't believe it; Bastian has gotten both of us jobs, just like that. I had been telling him how hard it was to find work out here, and he'd skooshed it two days in. I was thankful, if a little chagrined that a complete novice had outmanoeuvred me on the backpacker jobs front so quickly.

The warehouse was within walking distance from the hostel, and there were a lot of us from the hostel making up the numbers. It was straightforward stuff; just opening, transferring and repacking crates of CDs, and everyone got stuck in. I had only been there 30 minutes when my female tracking radar achieved a direct lock-on. There was an Asian girl at the other end of the room wearing tight fitting blue hipster jeans and a cut off yellow top. I was immediately struck by her beauty. She was sublime. But there was also something else that affected me ...her sadness. There was a definite sense of sadness around her; her posture hunched despite her prettiness, as if she was carrying a heavy burden, and she seemed preoccupied and disconnected. She was an instant mystery; why was such a lovely girl so sad?

I like mysteries.

I thought about the opportunity to make her feel happy.

She ended up working on the CD stack directly beside mine. I wracked my brain furiously trying to think up a good conversation starter. She was East Asian, but from which country I didn't know. My first thought was to try some of my newly acquired Korean...

...but then what if she wasn't Korean; would she take that as an insult?

I wavered and wavered...

and then...

...she was gone.

I mentally kicked myself over and over again; what on earth did I have to lose?

Lady luck smiled on me shortly after, as she started to work on one of the other stacks I'd already completed. I saw my opportunity and swept in;

"Sorry, I've done that one already"

"What?"

"I've done that one already"

"I don't understand"

"I'VE DONE THAT ONE ALREADY"

"What?"

"I-VE DO-NE TH-AT ONE AL-READY"

It turned out either her English or my accent or both were holding up the proceedings of communication. I had to get the warehouse boss to translate.

Obviously a little flummoxed by what had just transpired, she was sent off by the boss to work in another area.

So much for opportunity.

That had went about as bad as humanly possible.

I felt terrible. I swiftly drowned myself back in packing and unpacking in an effort to numb the post-mortem self-analysis. I couldn't even confide in Bastian, as he was at the other end of the hall. That didn't stop me from hearing him though, as his Teutonic motormouth was in full flow as usual, and carried over the din of a warehouse full of people. It sounded like he was having a good time.

A most pleasant surprise. At the end of the day Bastian was waiting for me at the exit. But he had someone waiting with him; the pretty girl. It turns out that I'd been listening from a distance to him chatting her up all afternoon. He'd asked her if she wanted to come hang out at our hostel and go to dinner afterwards, and she wasn't sold until I came along and Bastian changed his pitch to "let's all make new backpacker friends"; which worked. Any jealousy I had towards Bastian evaporated with the opportunity he'd just created for us to hang out with her.

Anna was her name and the more I spent time with her the more entranced with her I became. To go with her lovely figure and natural looks, she also had an effortlessly individualistic sense of style. It turned out she was Korean after all, and I got to show off my knowledge of Korean culture over dinner at the local Japanese restaurant. We ended up going for drinks at the bar in her hostel and I told her with no little enthusiasm about my big journey, which also seemed to go down well. We said goodnight and exchanged numbers with her, knowing that we'd be seeing each other the next day at the warehouse.

It had been a most fortuitous day.

01/September

At the end of CD packing day II, I was greeted by the sight of Bastian standing waiting for me with another gorgeous Asian girl as well as Anna. Mai from Japan was her name, and she was sharp witted and fluent in English as well as being absolutely lovely. You had to be impressed by Bastian. Mai had her own small company in Japan, which turned over so well she could afford to take long holidays. It was quite a resume for a petite girl that looked about eighteen, but was actually thirty years old. We all had dinner together back at the hostel; it was wonderful, and afterwards it became clear that Bastian was developing something of a crush on Mai. Beautiful, intelligent, a successful businesswoman who loved cars; I don't think he could have previously imagined meeting someone like her if he'd cut out a bunch of pictures and stuck them in a computer. The fact that she was also about the same height as him was just a bonus.

02/September

The CD work was finished for now, though more was rumoured in the coming weeks. We now had a group of friends to hang out with and the trip down to the hostel kitchen came with the thrill of anticipation of maybe bumping into one of them. Everyone, apart from Mai, was looking for work now, and I mixed in tours of the recruitment agencies with my beloved pastime of studying at the library. I returned to the hostel after a day of searching and studying to be greeted by:

"Hey man, I found a German girl. She's beautiful!"

This wasn't Bastian's only news. He'd gotten a job at a car showroom. Apparently he'd just waltzed in, started *chewing the fat*, and talked and talked until they gave him a job in the backroom cleaning cars. I had to take my hat off; Bastian made things happen. Everyone who travels to Australia, or any new and promised land, has a dream of landing on their feet as their best self; *transformed* — making contacts, creating opportunities, building a life from scratch in a whole new world. But most, something like myself, end up lurking around hostels, skint, for months on end. Bastian though had done it; a week in he'd got himself a job that wasn't even vacant just by following his passion.

Bastian introduced me to his other German friends; two young guys on a year out and the aforementioned beautiful girl. Brigit was tall, blue eyed and blonde haired, and she was more than just a pretty face; she was currently in Melbourne doing a work placement for a second degree in Medicine. Despite her exceptional qualities she didn't let them go to her head and she had a completely down to earth personality. I was more than impressed.

Taking the opportunities a Friday night and the Bastian-mobile represented, we lost the two boys and invited Brigit and Anna out for dinner at south beach. We had some excellent Italian food and wine and got to know each other better. The more I got to know Brigit the more I liked her; she was the same age as me and we had a lot in common. She liked *Smashing Pumpkins*. She was great to talk to, and talking to her also played into my cunning plan of trying to make Anna jealous. Later, we stopped by the beach to hang out and watch the southern hemisphere stars. What a wonderful evening.

03/September

I had gotten the ball rolling on one of my other long term plans by contacting Jeff, one of the worldwide network of instructors in my internal martial arts school, for lessons. He graciously agreed and I got the chance to join up with his private students. Although a *Taijiquan* teacher, he knew enough *Shaolin* to set me up in the basics. I was pleased to be going about kung fu the old school way; six months of demanding stance training (*zhan zhuang*) and leg stretches before learning even basic techniques. Jeff was a master in his own right and it was a great experience to learn from him.

04/September

One of the great things about being friends with girls is that you get to meet other girls. A new day at the hostel, a new arrival. Hye Jung from Korea and was yet another thirty year old that could pass for a late teenager. Her complexion was absolutely flawless and she shone with health; silken hair, alabaster skin and eyes bright. Together with her height, around six foot, she was quite the striking young women. But like the other girls, she also had a wonderful personality, soft and gentle — a true Korean princess. Through Mai we'd got to know a lot of the network of young Asian women at the hostel, and they were all delightful people. They all had different backgrounds as well; Sakura was a student from Hokkaido taking a year out, quiet and pretty, Cho, a tall and exceptionally sarcastic young student from Beijing, Borah, a trainee hairdresser from Seoul who had a love of musicals, and Tracy, a post-grad from Taiwan, who was even more of a garrulously outgoing socialiser than Bastian.

It was great to hang out with them, to chat about our homes and cultures, and share our plans and experience in Australia. But the best news of all was that Anna had gotten a job at an international hotel, and had moved into our hostel now. I suspected this was more to do with the company of the girl's than mine and Bastian's presence, but either way I was determined to make the most of the opportunity.

05/September

Despite the fact that he'd just landed a job in a car showroom, this hadn't slowed down Bastian a bit.

"Hey man, I've started my own business!"

What?

I couldn't believe it at first. But he'd spotted a niche after buying his own car and had put posters up around the local hostels advertising cheap old cars for sale. Of course, Bastian didn't have any cheap old cars, but he'd struck a deal with the used car dealership to provide a backpacker premium and a small cut of the profits on any cars sold through his channel. I wasn't sure of the feasibility of such an arrangement but his audacity and enthusiasm was as impressive as ever.

I was really enjoying life in this hostel, and I had a good routine of walking to the park to train *zhan zhuang* and *qigong* in the morning, then on to the library to study till the afternoon, then back to the hostel to shop for dinner, then to get showered and ready for making dinner with whoever was around. After dinner, back to the park for some more *qigong*, then back "home" to the hostel to hang out before bed. Knowing so many people meant that hanging out was infinitely more enjoyable than it had been back in Sydney; there was always someone to talk to or prepare a meal with, and we made sure to make new arrivals feel welcome.

06/September

Spending more and more time with Anna meant that I was falling for her more and more. I'd been in love before, and this was love again. But it was stronger, even stronger, than anything I'd felt in the past. As my mind wandered onto my future plans, for Tasmania, New Zealand and Fiji, all I could think about was how much I wanted her to travel with me. More than anything else in the world I wanted to know her and spend time with her; be with her, and travel with her. We'd only just met but I was giddy whenever I was in her company. But I couldn't travel with her as just a friend. I needed to make her my girlfriend.

I was leaving for Malaysia on the 28th of September for two weeks — would I be able to woo her in time? That gave me about two weeks to make a move and confess my feelings, which would be some sort of world record for me in the arena of romantic relationships.

Could I do it?

Would I fail?

How would I feel if I failed?

07/September

More CD work had come up. This time, however, it was out in the suburbs so a lot of the hostellers weren't able to take part. Thanks to the Bastian-mobil we had no such problems. Decent money for easy work. Australia was a golden land for Bastian, and he spent a lot of time grilling our employer about business opportunities. The boss, no doubt dealing with the rather tough bottom line that went with shifting excess stocks of CDs and the employment of backpackers, was more than a little chagrined by Bastian's relentless ideas and invincible self-belief. I put my brain in neutral and got on with packing shelves. But just like when trying to meditate in the park, I kept thinking of her. Whilst CD packing I would find myself listening to the radio that was on in the background, with my heart singing and soaring along with the sentiments of the corniest teeny bop. It was with some horror I realised what was happening. My *qigong* training had resulted in the development of...

...Emotions.

08/September

With two more days of wages burning a hole in our pocket, we planned a little party for the girls to go with dinner on Friday night. I found a candy store that imported UK sweets, including the assorted confectionary delights of my childhood and Scotland's second national drink; Irn Bru. Bastian sourced some proper imported German sausages. It went down well, the girls delighted by the treats, even the sickly sweet orange soft drink.

Personality is a funny thing. I noticed that Brigit and Hye Jung would not be at the same table together for any length of time. When Brigit wasn't around Hye Jung occasionally asked, "don't you think she (Brigit) is arrogant?" When Hye Jung wasn't around Brigit occasionally asked, "don't you think she (Hye Jung) is arrogant?" Two tall, intelligent, kind and beautiful young women; and the only people who couldn't appreciate their qualities were each other!

09/September

I often felt like pinching myself here in Melbourne. I was fulfilling my goal of daily *Shaolin* training with no pressure of time, there were great public resources for my vocation of writing and thanks to the work that was on the go I had enough money to live. I had met a great friend and I was spending every free leisure hour in the company of beautiful women from all corners of the earth. Life was good again.

I came back from my evening *Shaolin* training to find Bastian holed up in bed in our room. He pretended everything was normal but I could tell from the total air of despondency that surrounded him as well as the fresh tears that he'd just been crying. He'd tried it on with Mai and had been knocked back. I knew what a broken heart was like and felt for him. I hoped I wouldn't be in the same state in a week or two's time.

10/September

Myself, Bastian and another hosteller Bastian had bonded with over shared interest in computing went through the ritual all those passing through Melbourne must go through; to visit the Sky Deck of the Rialto Towers for the evening city view. As amazing as the night time cityscape was from the 55th floor, I wished I was seeing it with Anna. But the girls had already seen it and done it.

Even more amazing than the view was how life rich and exciting life had become since arriving here and meeting Bastian. Individually, both of us had social difficulties for different reasons; Bastian was too outgoing, without barriers, and thus people found him difficult to trust. My introverted nature on the other hand was difficult for strangers to trust for the opposite reason; too much of a barrier. But together we made quite the team — my yin to his yang creating a dynamic of creative action.

11/September

"I don't like Anna."

We had bumped into each other downstairs over breakfast and she had ignored me. The rebuff had cut me deep and I was fuming.

"Hey man, calm down. Maybe she didn't see you." counselled Bastian.

"She did see me."

"You're overreacting. Maybe she had something on her mind."

Bastian himself seemed to have shrugged off Mai's choice.

"There are plenty of chicks here in Melbourne — that's what we need to do man; we need to go out clubbing!"

He was a hard man to keep down. He was losing interest in the car showroom already and wanted to get involved in website development while he was out here. Remarkably though, he'd managed to sell a car already. We took a day out from the hostel routine and used the freedom a car entails to further explore Melbourne under a warm sun.

I saw Anna and the girls over dinner and there didn't seem to be any problem; with me, or with the group. After the afternoon of driving I had actually been looking forward to the chance to see her again. And I couldn't take Bastian's suggestion of going clubbing seriously while there was even the remotest chance she might be interested in me.

12/September

The price of bananas was always a shocker. Feeling the need for healthy food I had bought a few and it had cost me over $6. Hostellers wondered at my extravagance. I wondered at my extravagance; as I had just put them in my shopping bag without thinking. How can it cost more than $2 for one banana?

The reason I came to discover was Typhoon Larry. The very one that had hit Queensland just after I had arrived way back in March — it had wiped most of Australia's Banana crop. If a singular storm could cause such long lasting shortfall and effect on the price of living, what would a worldwide destabilised climate do to the economy and our lives?

13/September

I bumped into Anna in the internet section of the hostel. I showed her some photos of my big trip, and she reciprocated by showing me some photos she had online of her life; all the way back to childhood.

Afterwards, I had to go the room to be alone. I didn't want anyone to see me cry. The immensity of life had hit me like a thunderbolt. Looking at the pictures of little her back in the 80s, I had thought of little me back in the 80s. Both of us had looked up at the same blue sky overhead, never knowing of each other's existence or that we each looked up to the same blue sky. Two little lives lived out simultaneously in complete un-knowing; so different and yet the same. And now here we were — a chance temporary collision in a whole other part of the world.

Even if nothing came of my fledgling friendship with Anna; even if I did not achieve my goals of a love relationship with her and we parted ways forever in a few weeks only as mutual acquaintances, I knew that I would treasure those moments — those moments when she gave me a window onto her whole life — to my dying day.

14/September

Most of the girls had found jobs, so we used text messages to coordinate meeting up for meals. I had started dropping Anna playful little texts whilst I was in the library and she was at work. It was easier than flirting in person.

Then the bombshell:

Do you want to meet me for dinner?

I stared at my phone in disbelief. I never expected that. My mind started revving up, trying to guess all of the possible intentions behind the question, and all of the possible outcomes that a reply could result.

I replied:

Sure, what time? Should I pass on the message to Bastian?

The response:

I am asking you, not Bastian.

Holy Fuck!?

What did it mean? Was she asking me out for dinner? Had she already asked him along and didn't need me to pass on the message? Was I going to meet her and all the rest of the girls at the restaurant? Foreign languages and cultures mean that you can never be sure exactly what is meant, especially by a text message.

I met her from the hotel after work and we went for a Chinese in town. I was on tenterhooks to begin with, but loosened up as things went on. It turned out it was just the two of us, but I liked that; the chance to monopolise being with her, to stare into her eyes across a table and just talk. The conversation was good and fun, but light, and there seemed to be a lot she didn't want to share.

Later we walked across the bridge over the river and took in the evening view of the city. With the lights of the backing skyscrapers and the bustling bars of the riverside reflected in the water, it was sublime. I was in internal paroxysms trying to work out if now was the time to make a move, to make my feelings known, but then she startled me:

"When you talk about these places you want to go it's so exciting. I want to do something like that. I'd like to travel with you there".

She wanted to travel with me?

I felt like exploding. It had been my heart's deepest desire since pretty much when my eyes fell on her ...and now she actually wanted to go travelling with me!

But then the analysis — *what if she meant as friends? To travel together, just as friends?* The thought filled me with a base horror. To travel as a friend with a girl you have an unrequited passion for is possibly the world's most hideous form of torture.

Contemplating the two possibilities — "boyfriend", or "just friend" — I felt like I was balancing over an abyss.

We sat down on a bench at the riverside. And for the first time, she opened up to me. And then she started to cry. I'd hit a sore point with my questioning, and she broke down as she told me about the cause of her pain. She had been working in Australia for the past two years with her sister, but recently they had left the farm they had been working on after a major fall out. A fall out so bad that it had seen Anna travelling to Adelaide and then on to Melbourne by herself, and her sister flying out to the Philippines to travel on her own.

I didn't know where I stood with her, so I tried to comfort her as best I could without upsetting her personal boundaries. Her breaking down in tears after spending time alone with me for the first time certainly wasn't my planned outcome for the evening, but I felt closer to her now that she had opened up to me. I thought that even if nothing came of the relationship, and we both went our separate ways in a few weeks, that I was privileged to share this time and this conversation with her. I took her for a glass of wine — I could only afford one each — in one of the riverside cafes and we kept talking; she seemed brighter and brighter after getting the weight off her chest. We walked slowly back to the hostel along the waterfront and said good night in the lobby.

15/September

"Hey man, where did you guys go last night?"

I told him about having dinner and walking along the river together. He seemed a little put out at not being invited, but I was too busy being overjoyed with getting to spend time alone with her. I still couldn't work out if she liked me just as a friend or not, but either way I was both stunned and delighted with the progress in the relationship.

Things just kept getting better and better. The recruitment agency called to let me know there was more office work on the go. It was more nine to five 20 dollars per hour manna and as a five day placement, with possibilities for more in the coming months, it also synchronised with my upcoming travel schedule. Which was nice.

16/September

Saturday night was upon us and I knew this was the time I had to make a move. There was a comedy night on in the hostel, a proper one with professionals and all, and everyone was going to be there. I spent a lot of time getting ready. The place was a riot when I got downstairs, everyone was cooling off after the week of work, and the quality of performers had brought in young people from all around. There was a happy hour on and we ordered as much as possible.

I saw my opportunity. Anna had been left temporarily alone, without the male or female company that usually surrounded her when we were hanging out, on one of the big comfy couches in the venue room. She must have spent as much time getting ready as I had. She looked fantastic. She was wearing a short skirt that showed off her thighs and I barged into her playfully as I sat down. She responded, and pressed her thigh into mine as we talked. It was electric; her legs, those legs I'd been admiring for the past two weeks, were so lush, and now they were pushed into mine. As we didn't know each other so well, I had plenty of scope for making small talk.

I'm not the sort of person that can talk for hours on end, so I suggested we play some games; we took on Bastian and Mai at table football and had some fun. Later, the girls congregated in the seats to watch the stand up, and I stood and had some beers with Bastian and the German guys. It was really rather good, and one act was even a half-German, half-French raconteur. Bastian got pretty excited about this as he was also half-German and half-French and went up and introduced himself to the performer afterwards.

The guy was Australian. It had been a good act.

The students and assorted 'outsiders' who had came in for the comedy show started to filter away once the last act ends. I ask Anna if she wants to join us for a game of pool. She surprisingly turns out to be a bit of a shark; pool being the leisure of choice in the world of Mildura farms. I was pleased with the choice of pool — a chance to drink and play and talk, but not too much pressure either way. It meant her spending time with me rather than anyone else, but as time went on the conversation began to dry up, and it seemed more 'matey' than romantic.

The girls disappear to the toilet at one point and an English guy — who I'd seriously considered beating to death during the week, as he was always on to her — comes up to check on my progress:

"How you getting on with that Korean bird?"

"I thought I was doing alright earlier ...but now, not so well."

"You are mate! You are!"

"No, think I'm starting to crash and burn."

That's what it felt like. After the flirting and excitement of earlier in the evening I was now on the downslope of the alcohol buzz and the tiredness and the running out of topics of conversation. I didn't want to spoil our fledgling friendship by making a move at the wrong time, but there wasn't anything particularly romantic or intimate about our chatting in the venue hall. As the night was drawing to a close I felt like I was running out of time. Would thinks continue on next week with just this 'friendly' feel?

A bunch of folk, the bulk of the Anglo-Irish group were going clubbing and Bastian was joining with them. I started to get my hopes up again. Dancing sends signals, dancing gets you close, and I like to dance. But Anna had work in the hotel the next day and wasn't going. I decided not to go clubbing either, as I thought it'd send the wrong message to her.

There were a fair few regular hostellers not clubbing, so we all hung out in the kitchen and drank soft-drinks après bar closure. I was coming down and getting more tired, but just tried to keep it together and play it cool till the night came to its natural end. I ended up walking Anna to her room, as it was on the same floor as mine. The pressure started to build as I got closer to her door. *What was I going to do?* After our night by the river and our talking on the couch I had to tell her how I felt. I couldn't just let it end here. This might be my only chance. *But then what if she didn't like me like that? What if I'd read the situation wrong?*

She unlocked her door and turned to say goodbye.

I said it first:

"Well...goodnight."

She responded;

"...Goodnight."

And then the dams of internal conflict broke, and I did the only think I could think of to give her a message:

I leaned over and kissed her cheek.

And then...

...her forehead went in to a half-frown and she just kind of looked at me.

I staved off the onrushing crisis of what had just transpired by clamping down hard on myself and projecting what I hoped would be a flash of cool. Gave her a playful half-smile and waved as I turned away. Hoped I could write the whole kissing gesture off as a European-continental style courtesy.

As I turned the corner towards my room and out of sight the post mortem started. *What the fuck had I done?* That look. I had ruined everything. *What was I going to do next time I saw her?*

Bed was a sanctuary but no sleep came. I turned the events of the day over and over again in my mind. *What could I have done different? Why did I do what I did?* Most of all, I dreaded the next week and the effects of my kiss-rape on Anna and the whole group at the hostel. I had spoiled everything — between me and Anna, between me and all the girls, between me and Bastian and all the girls.

And then my phone went. It was fucking 2 am. Bastian?

I squinted at the lit up screen in the darkness.

It was Anna. *Holy fucking shit.*

Was she going to tell me she hated me? That I'd violated her with my kiss?

I needed to know, so I opened up the message.

If you are awake come and meet me by the elevator at 2:30 am

Holy fuck. *What did that mean?* Had I crossed her boundaries enough that she needed to complain to me in person? I couldn't work out the meaning of this mid-night text, so I decided to go and face the music. Better to resolve things now than in the morning.

She was standing by the elevator when I came round the corridor. I scanned her body language for any sense of upset. Her look and expression was totally non-committal; she didn't look pissed off, she didn't look happy.

"Hi."

"Hi."

"I got your text."

"I want to talk to you — come and sit down." she gestured to the seats.

Uh Oh.

I sat with her, but didn't want to predicate. I waited for her to talk.

"Andy... why did you kiss me?"

"I was just being friendly, did you mind?"

"No"

And then some more silence.

"Andy... there's something I need to say to you?"

I waited, wondering.

"Andy... I like you."

What a relief. But what did she mean? It felt like I was balancing upon a precipice of possibilities once again. My fear came up straight away.

"You mean... like a friend?"

"No, not like a friend."

I couldn't believe what was happening.

"You mean, like a ..."

I was afraid of failing at the final hurdle.

"...brother?"

The moment it was out of my mouth I was like; *what da fuck*? Why did I say that? Talk about wrenching defeat from the jaws of victory.

"No."

My brother comment had just about killed the moment. But she composed herself and went on.

"Would you like to go out with me?"

Disbelief turned to awe.

"You mean as a boyfriend?"

"Yes, as a boyfriend."

All my dreams had come true.

"Of course! ...I mean ...I didn't know if you liked me like that."

"I do like you like that."

I was holding her hand now. We were sitting together and looking at each other intently. I couldn't believe what was happening. I didn't kiss her there and then because I still felt like there was a barrier there. 9000 km and polar cultures.

"Do you want to go out with me tomorrow, after work?" I asked.

"Sure."

And now we had a date. A proper date. Just the two of us. I was delirious as I walked her back to her room. We got to the door unlocking, saying good-night stage again. I was a lot more confident this time. As our eyes lingered, I decided, *fuck cultural barriers*, this deal has to be sealed. I leaned over and kissed her.

And, oh, I was eighteen again.

It felt like the first time, all over again. A bit messy, a bit of teeth, but *oh... oh...so real*.

It was wonderful and I didn't want to stop. She was so accepting, and I pressed into her, my hips fitting the curve of her hips like a dream.

Dreams really can come true.

We didn't want to stop. Club goers started filtering back into the hostel and we started getting whoops of approval from passersby. We went into the girls toilets for privacy and kissed and kissed and kissed, and then when interrupted there, we went out to the seats in the hall. And that's all we did. Kissed. Kissed and kissed and kissed and kissed. It was heaven.

Eventually, we had to stop. She had work in the morning. I waved her off as she went to bed, and tread the corridor to my room once again, this time floating on unbelievable fortune. I turned along the corridor and out of sight of her room, and as I did I let go of the calm. I ran, hopped, skipped, jumped and punched the ceiling with a cry, then collapsed on the floor laughing and grinning in delight, overcome with the triumph.

Ecstasy.

17/September

I couldn't believe that I had pulled her. It was a miracle.

I saw her in the kitchen making breakfast with the other girls first thing. A quick kiss on the cheek cut through the morning-after glaciation.

Later, I happened into Bastian in the men's bathroom. Told him, whilst grinning ear to ear, that I'd made a girlfriend. He didn't seem impressed, in fact a little annoyed.

I was surprised at his response, as I thought he'd be happy for me.

I tried to put myself in his shoes; whilst I was riding the immortal wave of romantic triumph, he was experiencing the counter archetype of partner-in-crime getting stolen by girl.

18/September

Our first date was the science centre. We took a train from the main station. All the time I was with her I kept getting hit with waves of disbelief. *Was this really happening?* I couldn't believe it. This beautiful young woman, this girl I'd been fantasizing about constantly for weeks, was now with me. It was incredible.

I kept admiring her, and wondering if the moment was real.

She was magnificent, and to walk along holding hands with her was like a living dream.

We messed around in the science centre amongst the day-tripping school kids, but the exhibit wasn't anything special. The observatory on the other hand, was great and we sat back in the auditorium seats and held hands as the spectacle of the artificial heavens revolved.

After dinner we went to the cinema. *Silent Hill* turned out to be the worst movie ever made. I'm sure fledgling relationships have been killed by less worse movies, but if she could forgive my choice with that, she could forgive anything. Not the most inspired selection of activities for a first date, but just the sheer experience of being together with her made it. It was a blissful day.

20/September

Our long walk along the riverbank when she had first confided in me had turned out to be the first of many. After dinner with our hostel friends we'd go for a walk by ourselves along the banks of the Yarra River. It was a busy promenade even in the Southern Hemisphere winter, and Melbournites paraded in their fine winter wear. Not quite as well budgeted or equipped, we appreciated the heat from the pyrotechnic displays outside the Casino. Our lives became more and more entwined every day.

21/September

The office work rolled in. A whole bunch of temp staff had been recruited by an agency, and we just got on with what we were asked to do. It was a piece of cake and I enjoyed finally catching up with Bastian, Anna and the girls to join the world of work and earning money.

24/September

Things had moved on between myself and Anna. We'd caused a minor scandal which resulted in Anna being ejected from her room, but fortuitously, this allowed us to move in together to the same dormitory, which in such a large multi-storey hostel was one of many which were otherwise unoccupied. We'd talked about sharing a room but couldn't afford it, but a dorm to ourselves was just ideal.

Malaysia

28/September

Clear skies make for the perfect aerial experience. A spreading panorama of sprawling Melbourne gives way to the open countryside of Victoria. It was a veritable simulacra of Europe, all patchwork fields and tree-lined borders, but with one important difference. While Australian vegetation looked pretty lush when you are on the ground, from the air its desiccated nature becomes apparent. This Little England was a relief of dark olive tones.

The farmland gradually washes out in deepening tones of brown and green until bands of dark blue bush land emerge, then melt again into rolling dust. As we travel further across South Australia, fault lines rise and cross the barren landscape like the spines of great snakes.

I spend most of the journey enjoying action films on the movie channels. I got chatting to the Aussie woman sitting beside me over dinner and it turned out she was planning to become a WWOOF host. She had her own biodynamic vineyard and we discussed the WWOOF project. She hadn't had any guests so far and didn't know what to anticipate. I was equally curious about biodynamics in practice. Some of the methodology described was rather mystifying. Interestingly, her vineyard had been undergoing a comparative study with others in the region conducted by one of the higher education facilities in Melbourne. Apparently, despite distinct disadvantages in both location and environment, her wine outperformed other conventional vineyards in many aspects of the study — including being voted top for quality — and the grapes in her orchard ripened quicker than any other. It was yet more food for thought.

The flight slipped on and I opened the shutter as we passed over Kalimantan; Indonesian Borneo. A few stars flickered in the night sky, reflected on the ground by singular lights from sources unknown amongst the mirroring blackness. Throughout my travel experience stereotyped imaginative preconceptions had fallen away as I encountered new lands, to be replaced by vibrant living images of people and colour. No matter how many new places I had encountered, the darkness of unknown lands remained. My mind still said:

"Here be Dragons."

Truly, you never know unless you go.

29/September

I had checked into a hostel in Singapore for an overnight stay before continuing my journey north via Kuala Lumpur. It was nice to loosen up after the flight and soak in the sensation of being in Asia once more. The bus to Penang cost about ten dollars whereas a flight came in at a several hundred.

No contest.

The journey north was long and familiar. From Singapore to Kuala Lumpur the overwhelming sight being the refrain of Palm Oil plantations stretching off into to the horizon on both sides of the road. The air had been hazy in Singapore, and it was with some alarm that I noticed that it wasn't diminishing as the journey progressed; in fact it was becoming thicker. By the time the bus pulled into Kuala Lumpur visibility was reduced to near zero; a country-spanning smog.

I had a few hours to kill in KL, and I found out from one of the locals that the pollution was the consequence of forest fires drifting across hundreds of kilometres of land and sea from Borneo. The march of Palm Oil's dark harbinger. The journey continued under the pall, a pall that became reflective of my mood as I watched people going about business as usual, seemingly oblivious.

The bus dropped me off in a small town on the northwest coast. I checked into my room at the hotel and my eyes caught the sign in the bathroom:

"Save the Planet — Please minimise towel use."

What a sickener. SAVE THE PLANET? *Towels?!* I had to laugh, as a mental defence mechanism as much as anything else. Again, that contrast of lunatic detail with the apparently accepted truism that planetary salvation is required. I looked around at my surrounds and thought; *I'm sitting in what amounts to a very large European building plus air con in a town situated not very far from the equator.* Given its size, the dynamic activity within, and the climate and sunlight of the region, the building could be engineered to generate its own energy. Currently, it was just one giant fossil fuelled white elephant.

But for global ecological and humanitarian salvation, read: hotel managers wanting to minimise some utility costs.

06/October

The wedding courses were a six day series of complimentary courses followed by two dinners held by the bride and groom's families, respectively. My childhood *Shaolin Kung Fu* ambitions were realised at the first complimentary course, and although unable to follow it fully it was the perfect foundation for future training.

Taijiquan was also magnificent and extremely powerful. Grandmaster Wong drops pearls of wisdom almost constantly and his martial art teaching is something deeply existential. One point that struck me as he talked of the philosophic origins of *Taiji* and *Wuji* was a comment on original sin simply meaning:

"Separation from God."

The remarkable ability of the spiritually realised to transcend cultural barriers; illuminating concepts whatever one's background.

Qigong was my personal favourite: training with a Master is an exceptional experience that simply defies description to the uninitiated. We were introduced to some advanced exercises, including *Dan Tien* breathing and *Yi Jin Jing*. It was atomic. Truly extraordinary. I really felt the *qi* going to work; what is described in *qigong* terminology as "good pain"; the breaking through of blockages in my back and neck as the high power energy surged through my meridian system.

We had a lot of leisure time amongst the courses, mostly spent enjoying the climate, splashing about in the pool, or going for dinner in town. Many of the participants were British men younger than myself and I was incredibly impressed with their questing open-mindedness and fearlessness at such an age. They were much the same as young men everywhere, differentiated only in their relative moral incorruptibility. For example, their girlfriends would never have to worry about them being unfaithful.

Due to all the courses going on and the coming together of so many high level instructors, masters in their own right, there was an enormous amount of ambient cosmic energy in the hotel. I had the incredible experience, in my usual state of consciousness, of my hands vibrating with energy whilst I was having a cup of coffee in the restaurant. It was another startling observation: what on the outside amounted to just a gathering of people in an international hotel chain, the sort of thing that was happening all the time all over the world, from the inside was so much more.

07/October

The wedding celebration was an eclectic collection of spiritually open-minded people from all over the world and included people from every major religion, some from alternative religions, some of no religion. A section of humanity from Malay Muslim security guards to Irish Catholic housewives. The highlight of the bride's dinner was the spectacle of a German *Shaolin* master leading a conga of random international people around a banqueting hall mostly full of Malaysian Chinese to the beat of a classic 90's pop/dance anthem. Certainly not the usual preconception of what a "spiritual" gathering would entail!

08/October

As I travelled back down to Singapore I reflected on my time in Malaysia. Overall it had been a period of massive contrast. Spiritually uplifting and extraordinary *versus* environmentally shocking and harrowing. It seemed that only one eye was open.

As I looked out of the bus window at the blanket of smog that had stretched from Indonesian Borneo to here, the furthest northwest of the Malay peninsula, I couldn't help but think about the larger picture. All these people going about their lives, seemingly regardless of the visible shadow that overhung like the Sword of Damocles.

And when I turned my attention from the sky to the ground, there it was; coast-to-coast monoculture. Money has always grown from plants; now literally on trees. It was an emotional sight. Last year I had spent a night in a hide in the Malay jungles of Taman Negara. I had felt, saw and smelt the incredible richness of flora and fauna, the beauty and majesty of life in the forests of this land. I had heard the midnight symphony of the rainforest. Memories resurfaced; the rumblings of elephants, the glow of fireflies, the exquisite plumage of canopy passerines, the prancing espionage of a jungle cat, the thundering wing beats of a hornbill.

But what was I seeing in front of me now?

Not this, anything but this.

I started to daydream. All of the solutions I had been studying or had witnessed bubbled up. From indigenous agriculture and the structural mimicry of rainforest, to the working examples of the organic farms of the WWOOF movement; ideas spun off in my mind. As I looked out at the landscape, I could only think of the vast potentials in such a biologically rich land for sustainable forestry and a sustainable mixed agriculture. The human side of the country set my imagination off to the same degree; I thought about the massive potentials of passive solar, solar electricity and hot water; the ecodesign of buildings; biofuels from waste.

But all these things just surface dreams of mine.

How deep does the problem go?

I think it's all about **Value** in the end.

Only the Malaysian people could choose that. How they valued their own environment and how they integrated those values with the values of the global economic system to which their lives and livelihoods were now inextricably linked.

Return to Melbourne

10/October

Niggling at the back of my mind throughout my time in Malaysia was my relationship with Anna. I didn't feel it was at the properly cemented stage. She had other suitors, and I worried about what could happen in my absence with a young and attractive woman in a party-atmosphere hostel full of men who were younger and better looking than me.

Paranoia.

I couldn't wait to see her when I got into Melbourne. As soon as I got into the room the paranoia immediately splurged:

"Are you still my girlfriend?"

"*Yes!?* Are you still my boyfriend?"

I kissed her and it felt great to be back.

Better still, she wanted to talk about our travel plans together. Rather than take a holiday from her work in the hotel so she could travel with me for a while, she was going to quit her job completely so that she could spend as much time as possible with me before I left Australia.

It was the best news I could ever have had.

Her sister had arrived in the hostel during the time I'd been away, a reconciliation bringing them back together. I was overjoyed with this turn of events for Anna, and I was both nervous and excited about this early introduction to her family. Kelly turned out to be lovely, and a different personality to her sister. Older than Anna and ages with me; she was mature, with first impressions of quite a straight and serious demeanour. Underneath the business-like exterior however she was a gentle person; sensitive and bright, with excellent English. She had already settled in with the girls in the hostel, and Anna had even managed to get her a job at the hotel.

Being back at the hostel was like returning home. All the same friends and faces were there. Travelling in Australia is such an expensive business that people can easily spend months on end working just to live, never mind generating funds to go places. This meant that there were always some long term residents in the average hostel, and this made for the formation of some great relationships. It was great to see all the same clubs and cliques from two weeks previously, and all still getting on as well as ever.

11/October

I had the chance of more office work, which I had to turn down because it clashed with my Tasmania plans. So, whilst everyone worked during the week, I went to the work of continuing my research in the public library.

14/October

The weekend was another opportunity to party. It was Tracy's birthday and she had invited all the ladies for a girl's night out. It was with mixed emotions that I had to watch my beautiful girlfriend get dressed and dolled up properly for the first time in our relationship, only for her to go out on the town without me. Bastian and I made the most of our night of freedom to crawl the bars of the other hostels in the city.

I'd felt a little guilty going from spending almost all my free time with Bastian, to almost all my free time with Anna. But in the end time meant everything. I didn't know how long we had together and I wanted to spend every single moment I could with Anna.

Me and Bastian were still in the same hostel amongst the same bunch of friends, we just had fewer opportunities to "team up" and explore as we were doing tonight.

18/October

Anna and Kelly worked through the week, so I took the opportunity to join Bastian on a bus tour of the Great Ocean Road. She couldn't have afforded it anyway as she was saving for Tasmania, but encouraged me to go along with Bastian. The trip was better than I anticipated; the tour was well paced and included everything from forest walks to exploring coastal coves. And as the bus chugged along, the broken coastline of eroded Limestone pillars was stunning.

Whilst Bastian was busy chatting up some tourist, I found myself looking out at the expanse of the Southern Ocean, feeling guilty that Anna was going to the extent of quitting her job to come with me to Tasmania. But then I realised — continual harvest work for two years was not life. Making the decision to travel for travels sake; to take the risk and make the most of the opportunity to explore somewhere completely new and unknown while here in Australia; would be as good for her as it would be for me.

22/October

It was Sunday, her day off, and so a chance to experience the beautiful parks and gardens of the city. In the evening another dream came true; I took her to the top of the Rialto Towers, just the two of us.

23/October

I woke from a strange dream of spinning colours and darkness in which I was starting to become aware I was dreaming. I turned over to my girlfriend, a reflex embrace to wake her for some fun. She lent up on the bed and I saw a changed image: now an archetype of an Asian woman but half human, half machine. The bed posts dissolved and the bed began to float across the room, the walls around the periphery of my vision changing and shifting like liquid. The cyborg closed in on me, attempting a kiss of its own. I recoiled automatically from the monster. I was now aware I was still dreaming but it was completely lucid. I could not escape and its face touched mine. I became completely paralysed as it started leaching life force from me. I screamed. I knew it was a dream but I couldn't wake up. I couldn't move. I continued to scream and scream and scream and scream.

andy

Andy

"Andy?"

Sharp breath in as sunlight hit my face, heart racing. The room was no longer moving and my open eyes found my real girlfriend regarding me with a half-amused, half-concerned expression.

"You were going *mmnnhhnmnhmm, mnhmmnhm, mnhnnhmmm, mnnhmmm*"

"I was screaming in a dream but couldn't wake up!"

"What happened?"

"You turned into a robot and tried to eat me!"

She giggled.

"It's not funny!" I continued.

"Are you okay?"

"No."

I reflected on the experience.

"Thanks for waking me, by the way."

The adrenalin faded quickly and the memory returned and dissolved in decreasingly-vivid cycles throughout the day.

24/October

A random text message bringing a new surprise. It was Won *dongseng*. He'd made it all the way to Melbourne and was looking to meet up. I almost didn't recognise him when he showed up outside the station. He'd grown his hair. He looked great in fact, much more relaxed and at ease with himself and with life than he'd been in Brissie.

I invited him and his friend, another young Korean guy to the hostel and they got to meet Anna and Kelly. It was great to see him again, especially with him looking so confident and happy. He was still as sardonic as ever though, as he informed me of the comings and goings since I left Brisbane. Apparently Hwani had joined Noona on a farm for harvest work and Noona had left to go somewhere else shortly after. Won considered these connected events. Won wasn't in Melbourne long, as he was feeling the pinch and about to go looking for harvest work as well. It was fantastic to meet up again, and to see that things had been turning out well in Australia for all the other guys, not just me.

Tasmania

25/October

Excited as ever at the thought of somewhere new. Not only this, Anna was coming with me and all our time together had felt so natural and so good. Special lady, new places. Can things get any better?

Glimpses of Melbourne, then coastline through broken clouds, then sea, then the coastline of Tasmania. The flight passes extremely quickly and the plane begins to descend before I've even had time to finish my cup of midflight tea. The cloud cover is less thick over Tasmania and it looks similar to the Victorian countryside from the air albeit more hilly and wooded. Mountains jut through the clouds on the western side of the island and flashes of sun-reflecting lakes are apparent. We pass below the clouds as we make our approach to Hobart airport, the descent unveiling a wonderful view of spreading coastal channel, wooded hills and fields and familiar Australian-style houses. Most noticeable of all though is the towering vertical mount that rises abruptly at the edge of the city, dominating the panorama. It's an epic sight to greet the Tasmania tourist.

We check into a hostel and then explore Hobart on a trip to get supplies. Anna enjoys the quaint surrounds; post colonial bay-windowed sandstone houses sit amongst more modern townhouses, suburbs in the distance run into wooded hills that rise up around the small town. But by far the most powerful aspect of scenery was the peak to the North, much more massive from this angle than when seen from the air. The place has a distinctly different look and feel to flat, hot Melbourne. There's a stiff breeze, which has a chilly bite to it, cold and clean. A kiss from Antarctica.

26/October

A day of getting to know Hobart town. It is eerily quiet, with few cars on the road or people on the streets. Apart from being a ghost-town, it was almost like arriving home for me. From the police boxes, to the town planning, to the ornate stone fronts of Victorian era buildings, I could be in Scotland. We happen across a bric-a-brac shop that is quite the time capsule. On entering, I felt like I'd been teleported back to the early 80's. I hadn't seen stuff like this for years, and there were all sorts; comic books, antiques, furniture. Even to my inexpert eye I could see that a lot of this would fetch a pretty penny back home. The LP collection was out of this world; with such gems as an original *Ziggy Stardust* vinyl. But it was the amount of sentimental collectable regalia that if anything, made this little corner of Australia more British than Britain.

We arrange our first WWOOF placement together over the phone. Anna isn't quite one hundred per cent on the concept so we just make it a short weekend stay to get a feel for the arrangement. In the evening chill out in the hostels main room. *Waterworld*, the first Hollywood movie with a global warming theme, is on the TV in the background. Another thing we have in common: we're both not fans of the film.

27/October

I'd been wanting to see Al Gore's film for a while so with not much else to do we went to the cinema. *An Inconvenient Truth* a lot better than the sad Mad Max clone of the previous evening. I thought it an important film and a nice summation of global warming. However, I felt it left more questions than answers. Voting for Al Gore or anyone else will not save the world.

In the afternoon we visit the Hobart museum. The collection pretty limited due to the small town nature of the place, but a nice way to spend time nonetheless. The lasting memory provided by the haunting original footage of Thylacines from Hobart zoo; a hypnotic loop of film that is the last motive impression of their existence upon the face of the earth.

The D'Entrecasteaux

O ur lift is late turning up and the woman who picks us up much younger than expected. An English lady, she introduces herself as Rachel's daughter. Rachel can't pick us up because she is snowed in!

Graciously, Rachel's daughter and her husband provide us with a lift to our destination. Mt Wellington, that epic peak so noticeable from first arrival, is snow speckled as we head west along the Derwent river and down the coast of the d'Entrecasteaux channel. They remark that only two weeks ago they had bushfires near the suburbs, now it was snowing by the sea! The coastal drive a delight; rolling pastoral countryside interspersed with thick woodland of temperate deciduous trees. The freak weather had dispersed and now scattered white clouds drifted across a bright blue sky. The beautiful waters of the channel recalled the sky in an even deeper darker blue.

We arrive at the Overkamp household and are welcomed by their two excitable dogs. Jorg is an instantly friendly large man with a continental European accent, Rachel quintessentially English in voice and manner. Arriving like this feels a little like gate crashing a weekend family get-together, but they make us feel very welcome. We have a brilliant meal of tea and cakes and scones. Australia being Australia the topic of water is not far away from dinner table discussion. The snow is something of a blessing here as the Overkamp's have no mains supply and last night's precipitation a much needed boon for their rainwater tanks, which were approaching empty.

After Rachel's daughter and family leave Jorg shows us around their garden. It's relatively large compared to their bungalow, but not outsized. We're introduced to some banana passion fruit — a strange breed of plant, tasty yet very tart. It begins to rain and garden work is cancelled for the duration, a delicious lunch of oysters making a pleasant alternative to work. Later, Rachel takes us out to the shops, a nice little place with a great range of local organic produce. She takes her time on the way back, giving us a mini tour of the local points of interest.

Afterwards we help Jorg with dinner. He is quite a remarkable character, from Holland originally but having lived most of his life in Australia. By
nature a compulsive joker and storyteller, possessed of an unstoppable humour
and good cheer. His favourite topic of fun was himself, especially his recent
illnesses. After a recent operation for a kidney tumour (thankfully non can-
cerous on biopsy) he remarked that perhaps the doctors were trying to make a
"new woman"; as the operation had necessitated the removal of one of his ribs!

We got to know Rachel and Jorg better as the day passed. They both had
children from previous marriages. Jorg was an artist by nature and a handy-
man by profession, having constructed large sections of their house himself.
Rachel was a nurse, and came from a background in England which Jorg playfully described as "posh". We spent the evening relaxing, chatting and watching TV, which was full of much the same programming as back home in the
UK. It had been lovely to meet Jorg and Rachel, and we were feeling welcomed
as part of their household even on this very first day.

29/October

Every meal produced by Rachel was one to look forward to, a breakfast of
omelette and toast as much as anything. Our WWOOF tasks for today are
digging round fruit trees and putting down paper, manure and straw as mulch.
Rachel helped out intermittently but Jorg was banned from gardening due to
his health problems. The weather was pleasant and the day passed at a nice
pace. At one point Anna and I took a break from our work just to appreciate the surrounds together. The garden of the house dipped low, down to a
stream which separated it from fields for dairy cattle, and the surrounding
sight in the hollow was of hills of breeze blown green grass against sky. It could
have been anywhere in the world, just a small slice of life; grass and sky, but it
was extremely beautiful. I thought of how I could never have imagined this
moment before now, of how I could never have imagined the amazing young
woman I was with before she came into my life. I may only have been standing
in someone's garden, in Tasmania, mulching, but it was wonderful.

For lunch we had a tomato and pasta dish, continuing the exceedingly
good food theme that had started on our arrival here. After we finished our

afternoons mulching, Anna helped out Rachel with tonight's supper: fish pie. She has never heard of fish pie before and is very excited by the concept. Her willingness to help also means that I can escape to watch TV with Jorg. His ability to turn any occurrence into a joke or story is remarkable.

The fish pie arrives and is superb. To me, it is home cooking reminiscent of big family meals when grandparents still lived, to Anna, a culinary interplanetary adventure. Jorg understates considerably when he remarks that Rachel is an excellent cook.

30/October

It was our final full day at the Overkamp's and we continued our mulching after breakfast and began planting some fruit dreams down by the stream. A neighbour pops by and we have a long morning coffee break. The Overkamp's are delightfully relaxed people, and it's easy to imagine them getting on well with everyone. Once the guest leaves they comment on how much help they have had in recent times when Jorg's health has not been good, with neighbours constantly popping by uninvited to lend a helping hand. It was lovely to hear and easy to understand.

In the afternoon they surprise us with a complementary tour of the locale. We take a long drive along the Huon river valley in their campervan. It's a beautiful tour and the scenery of hills and deep, wide water — a touch of wilderness not so far from town life — is reminiscent of Scotland. But the *gestalt* of climate, rock and vegetation make it uniquely Tasmanian. We drive the length of the river down to the wide channel that connects it to the sea. We then turn inland before traversing over a steep hill and *voila*; we are looking out to the sea view of the d'Entrecasteaux channel with the Overkamp's house nestling in the village below.

Thai chicken curry is our evening meal, followed by Apple crumble for dessert. Anna has an incredibly sweet tooth and is in heaven with this new confection discovery. We chat about travel and the Overkamp's tell us of their adventurous caravanning trips around Australia. They were not "alternative people" like many other WWOOF hosts (indeed "alternative people" were one of the many topics of fun in the Overkamp household!), they were an elderly couple who liked organic food and who were just very open to the universe and what life had to bring. A lovely pair of old souls, comfortably relaxed in life and dealing with the health problems that can come with age with a sense of humour and fun, it was a real gift to meet them and to be made welcome in their home.

31/October

Breakfast at the Overkamp's for the last time. Rachel has made us French toast and it's excellent as ever. We sign the WWOOF guestbook before we leave and they go out of their way to drive us down to the bus stop and wave us off. Despite the short time we feel we've got to know them. The WWOOF stay has been a delight and a privilege.

The bus journey passes pleasantly through the countryside and we reflect on our time at the Overkamp's. I had enjoyed myself but Anna less so; she had found the mix of Scottish, English and Dutch accents at meal times difficult to follow. But she still wanted to try out other WWOOF places later on in our trip.

We checked back into our earlier hostel and spent some more time exploring Hobart. The original colonial waterfront — *Salamanca Place* — remained, now tastefully modernised with restaurants and bars. We did some planning; I wanted to visit the forest of the River Styx. Home of some of the world's tallest, oldest and most magnificent living creatures, or very large piles of woodchips, depending on which way you want to look at them. But car hire was expensive, so we'd have to forgo the forest and do things by public transport instead.

In the evening, the spring Antarctic chill meant that hanging out in the old hostel building in front of a real log fire was the place to be.

01/November

A trip to the Botanic gardens. Every city in Australia seems to have a Botanic gardens and thanks to the climate and locale they are usually very attractive. It was a long walk to the Hobart gardens, situated a few miles from the city centre. The walk was pleasant, a war memorial on a small rise giving a panoramic view of the channel, the estuary and the mountain-backed city behind. The gardens themselves were wonderful; rolling green lawns surrounded by walls lined by large trees. Miniature gardens and exhibits surrounded the main lawns and we strolled around and took it all in. Later we relaxed in the sunshine, which was warm enough to compensate for the perpetual Southern Ocean breeze. It was a lovely day and our relationship deepened as we talked about our lives; sharing sorrows; sharing hopes; sharing dreams; sharing joys.

02/November

On Rachel and Jorg's advice we took a bus to the historic town of Richmond. The scenery changes markedly as we pass beyond the hills of the Derwent valley; dry grassy plains and low rolling hills feature as we travel inland. The bus drops us off and we look about to see if we are in the right place — it's just so small, little more than a street with a few houses in the middle of countryside.

It's exceedingly picturesque; the main street features old colonial inns and 19th century sandstone townhouses, the buildings basically identical to examples from the same period in my homeland, a true relic of the colonial heritage. The small side streets are lined by trees and speckled by the Georgian-era houses, and terminate directly in the fields of windblown grass. The view all around the town is just countryside and low hills, and the combination of almost exact similitude but yet subtle difference from home, such as the surrounds of deep golden grassland flowing in the breeze, gives the hamlet to me the air of a dreamscape.

There are a lot of little tourist attractions peppered around the village including a mini-Hobart attraction; a little model of Hobart in the 1770s, like a Legoland set minus the Lego. Watching Anna enjoying exploring the exhibit I was awestruck by my own luck. Where else would I rather be? *Nowhere.* Who else would I rather be with? *No one.* Beyonce, Shakira, Jennifer Lopez — I wouldn't have chosen any fantasy figure in the public eye, or anyone I knew in person, over her.

We stop for lunch in the park and get swarmed by hungry and extremely well fed ducks and geese. Later, we take a stroll down by the riverside; the river lies in a small valley that cuts through the middle of town and its shallow depths, full of weeds and baby ducklings, make for the perfect centrepiece to the postcard. The sandstone bridge of the main road arches in a high parabola over the river and the view of this, the willow-lined banks and the little church behind fading into rolling country hills is the iconic shot of the little colonial town.

The town Gaol, now a museum, is a living link to the past. A curious aside is that the historical man behind the literature legend of Dickens's "Fagin" was once incarcerated here. It's easy to picture a Victorian character in this place, with an accompanying sense of impossible remoteness from London. The solitary cells in the gaol cast a lingering depression and the tree in the courtyard carries an extra touch of pathos; planted by a Tasmanian Aboriginal boy now long dead, as is his race.

Before we go I treat Anna to some scones with strawberry jam, which she more than enjoys. I can't take her to Scotland but Richmond is about as good an analogue as you can get 10000 miles away.

03/November

We journey to Launceston in the North of Tasmania. I spend the journey mostly watching the passing scenery, Anna does the East Asian thing of falling asleep in moving vehicles. Inland east Tasmania is much drier than I expected and less forested; mostly golden and brown grasslands and bare hills. I spot some of the culprits of the deforestation later in the journey; sheep, lots of.

Bony clusters of dead white trees occasionally appear in view, a reminder of the dryness of Australia. One section of the journey is dominated by the march of large sylvan skeletons across a valley floor. The scenery becomes gradually greener, more wooded and arable as we draw closer to Launceston. High land rises in the West and an escarpment comes into view in the East, creating an impressive frame for the plains and gentle rolling hills of the North spreading out in front.

We arrive in Launceston and have difficulty finding our Hostel, necessitating the help of a local taxi driver. The "how's the weather?" conversations in Australia are a little different to those elsewhere in the world and I get the perfect example of this from our driver, who breaks into a discussion of the paleo-climate of Tasmania without any prompting on my part.

Apparently this year is the driest on record, last year the wettest. This year likely to be an *El Nino* event. The water table of Tasmania was 250 metres lower 12,000 years ago and there was an ice cap on the high ground; a timescale which was little more than "the geological blink of an eye".

I could never have previously imagined having an unprompted casual discussion about paleo-climate with anyone, never mind a taxi driver in Tasmania. I say discussion, but monologue was a better word as there was no chance of me getting a word in.

What he seemed to be saying was this:

Yes, the climate is changing. There is now simply no doubt about that. But looked at from the larger scales the climate is always changing. I won't mention human activities because they have nothing to do with this. The fact that I drive a car for a living and cannot separate my daily existence and livelihood from the filling up of my fuel tank has no bearing on this soliloquy of mine.

04/November

A day of exploring Launceston; a particularly nondescript grid of streets abutting a river, overlooked by suburban dwellings on a hill on the west side of the town. We wander about looking to find the Launceston gorge, and surprisingly find it nestling behind that hill on the west side. The gorge has the primeval look of its brothers in the Northern territory, yet had been smothered and tamed by the Victorian settlers, its water flow reduced to a trickle by hydropower works. After my experiences in *Kakadu* I found it a strange place of contrasts; a modern public park with Victorian-era gardens and peafowl where once there was wilderness and raging torrents. How different would it have felt 500 years ago?

Spend the evening just relaxing in the hostel. Not so much to do in this one but there is a piano and Anna plays some classical tunes quite brilliantly considering she hasn't touched a piano in years. I never knew she played and it's another wonderful moment in the journey of discovery that is our relationship.

05/November

Our journey North coincidentally coincided with a craft fair in Deloraine, another historic Tasmanian town, so we take the opportunity to visit. The bus trip along hedgerow lined roads through the green rolling countryside in this part of Tasmania a pretty simulacra of a jaunt through the rural south of England. Anna loved it, and she loved the fair even more. There were all sorts of fun and games, brilliant arts and crafts and some stunning produce, with an Australasian twist; kangaroos, wallabies, allsorts on the menu. The native cuisine accompanied by locally brewed beer made for a great lunch.

Lorien — Part I: The DeClare's

John, our host, picks us up from Devonport. A gentlemanly fellow in late middle age he has a nice ambiance about him and he takes us for coffee in town before beginning the drive to Lorien. We travel through the pretty fields of Northern Tasmania towards the mountains and John narrates the landscape from an environmental and social perspective as we go. I don't think I've ever heard someone speak as knowledgably and as passionately about such issues before. His opinion on *Eucalyptus Nitens*, the genetically modified plantation trees which increasingly feature in the Tasmanian landscape especially vitriolic. It's great to meet someone who views life from a similar perspective but who can express and communicate those views so well and with such fluency.

The road to Lorien crests up over a high ridge and we stop to admire the views from the top. Mountains stretch west, south and east and to the north the rolling plains fade towards the sea. John is a wellspring of knowledge and talks about the history, the present and the possible futures of the region before we continue on. Lorien itself is situated on the eastern slope of a deep hydro valley, and the antiquated road that leads to it is narrow, un-surfaced and bordered by a steep drop that would be extremely dangerous were it not the exceedingly densely packed trees that would make it impossible for anything to roll down the cliff. The road is tough on cars but that's just the way that many of the residents like it, as the calibre of the road is the only thing stopping the large areas of forest that remain on this side of the valley from being logged out and replaced by plantations.

After passing along the steep, dark slopes the road begins to winds through open tree-lined fields and we soon come to the DeClare's residence. It's a spectacular place, self-designed and constructed by the DeClare's with various helpers (WWOOFers and other Lorien residents). A large two storey building made from wood with thatched straw walls, it is extremely roomy and comfortable; the open plan interior is airy and light. The house had been created exactly to their specifications and another nice touch was that it had no interior doors; only open doorways that were framed by the limbs of trees, enhancing the organic feel.

We meet John's wife Gemima for the first time, an upbeat and friendly woman a little younger than John, whose eyes convey a keen intelligence. We all get to know one another over tea and lunch on their front patio. The place gives a spectacular view of the north side of the valley; steep mountainsides clothed in the deep blue of dense evergreen forest. Our hosts were very frank and knowledgeable people and I got the feeling I was going to enjoy our time in the valley.

After lunch, Gemima guts and skins an opossum which had been caught in a trap in one of their vegetable patches. In the spirit of self-sufficiency the luckless interloper will be tomorrow's dinner. The carcasses of skinned decapitated marsupials have a pseudo-human look about them, and I start getting vegetarian thoughts at the sight. Getting up close and personal with the slaughter of an animal is a rare thing even though our daily lives are based upon it.

Many hosts have extra buildings to accommodate their WWOOFers, and this is the case on the DeClare's property. Gemima walks us over to their 'WWOOF house', an original farmstead building on the far side of one of their fields. We meet our next door neighbours on the way; the family's three dairy cows. It's a very old little cottage, an original farmstead, but not too run down and it's good to have the privacy of our own space to ourselves. Two bedrooms, a main room with wood fire, a kitchen and an outhouse bathroom with solar hot water, all electrified by solar power from their grid of solar tracking installations, make for a cute little dwelling. Our first job is to fix up the beds and clean up the place to our needs. We unpack and get settled in before returning to the main house for dinner.

We gather some salad for dinner from their vegetable patch and get introduced to multitudes of fluffy animals including chooks (chickens), rabbits, cats and guinea pigs. The rabbits and guinea pigs were originally part of an experiment on sustainable living, but turned out to be just too adorable to eat! We have a tasty omelette and potato dish for dinner accompanied by some exceptional imported wine, John being one of life's connoisseurs.

It's been a great first day, and the DeClare's give us some water bottles and torches to make our way back along the country road and across the field to the WWOOF house.

The air is cold enough to elicit a shock when we step outside, but this shock is overwhelmed by the shock of the night sky; an impossible sea of luminous gems, the *colours* incredible. The sky so clear that I couldn't even make out constellations; every patch of the great vault resolving into infinitesimal stars. Most were not even white; it really was as if an infinite amount of bright jewels had been scattered across the heavens. Red, yellow, green, gold, blue, violet. Astounding. Shivering, we gazed up in awe at the spectacle, watching little satellites wander across, the occasional flash of a meteorite.

Eventually, we began the walk back to the WWOOF house. What had been so bright and comfortable a few hours ago was now pitch black and sinister under the complete darkness of the trees. Our senses strained as unexpected sounds crashed and black shapes darted as we moved along. Anna got frightened and clung on to me; I played the masculine role and did my best to pretend I wasn't as frightened as she was! The shapes resolved under torchlight into bounding wallabies and rustling possums, but that didn't stop my heart from startling every time something suddenly raced towards us at speed out of the blackness. The whole place was completely *alive*. We occasionally stopped on the way to stand and gaze at the stars again. I had never seen a night sky so beautiful, not even in the centre of the outback. The combination of altitude and the cold, still air, along with the location of a remote place, deep in the Southern Hemisphere, creating the prefect stargazing experience.

We were completely freezing by the time we got into the house, and jumped with our hot water bottles straight into bed. Even with as many layers as possible we were still shivering and had to snuggle completely together to use body heat as warmth.

Freezing cold, in a shack, up a mountain, in Tasmania.

It was indescribably brilliant.

08/November

In the morning we get to milk cows, by hand, for the first time in our lives. Weird but fun. The raw milk is also a new experience, though not an unpleasant one; warm, creamy and sweet. Gemima then introduces us to their organic beef cattle herd. She is very knowledgeable and talks us through many details of how they raise and look after their animals. We transfer the herd to a new cell for grazing and Gemima explains the different types of feed crops used. Due to the drought the grass is not replenishing as quickly as it should and if the trend continues it will become increasingly difficult to maintain the herd. She was clearly deeply upset with this state of affairs but putting a stoic face on it all, and for the first time I was aware of how much energy farmers put into their livelihood and the bond they have with their livestock. There had been many stories on the news in Melbourne relaying the depression and devastation of farmers hit by drought across the county and I was very struck by the fact that it doesn't matter the political affiliations or philosophical inclinations of those farming the land: climate effects all.

Our WWOOFer task for the week was the weeding and hoeing out an overgrown hothouse. We got stuck in and it proved to be, as you would expect with working in an Australian hothouse, very hot and hard work. After a lunch of sandwiches we switched to moving compost in the afternoon, as the hothouse became a no go area in the high heat of the day.

For dinner, we joined the family for the opossum casserole. It had been slow cooked in fine wine for hours in the stove and turned out to be absolutely delicious; like eating turkey but with a slightly gamey tint. Every morsel melted in the mouth and I forgot what it was I was actually eating. Our dinner time discussion was equally as enjoyable; the family were very well informed people, extremely on the ball, and very enthusiastic proselytisers of their causes. They had quite a collection of news and research, and we spend the evening watching a documentary on *Global Dimming*; yet another scary concept to factor into the big picture. Later, we fight our way through hordes of wallabies on the way back to the WWOOF house, and this time light the wood fire to heat the house before retiring.

09/November

We have a basic routine for each day: clean the kitchen, feed the animals and then get on with our hothouse task. The family are attending an AGM in Devonport today, challenging the mooted expansion of a timber mill on environmental grounds. Over breakfast, we happen into Alice, our future WWOOF host from the family up the valley, and one of her current WWOOFers. They were popping by looking for Gemima, and seem very chilled out people. It's good to make their acquaintance prior to going to work at their house.

We get on with our scheduled work unsupervised; turning over soil and sifting compost in the gardens before vacuuming and cleaning the public floors of the house. It was the sort of house that was a pleasure to clean; a largely open plan ground floor, a second storey of living rooms arranged around a central void that looked down into the storey below, and a small third storey in the apex of the pyramidal roof. Light flooded in from the skylights at the top of the roof, and the third story functioned as a drying area for grain or rice. An antique wooden table and chairs sat in the middle of the house, lit from skylights in the roof above, beside a vintage wood powered boiler. A unique self-design by the family that reminded me of traditional dwellings in South East Asia, crossed with an old world country house. Luxury and sustainability are words that are not usually bedfellows, but were fittingly descriptive at the DeClare's.

After lunch we take a walk around the valley and check out the local community centre in order to gain access to the beloved internet and the rest of the world. Lorien really is a beautiful place; the dam-bounded valley forms a natural ellipse, protected on all sides. High peaks surround the three sides of the valley, with the fourth being the barrier of the dam itself. Nestled in this remoteness, our chalet rests in a field on a rise overlooking the lake, whose shifting levels are indicated by its border of flooded forest. On the opposite side of the lake, the valley floor rises steeply and is covered by a patchwork of pasture as well as natural and plantation trees. On this side rolling arable and pasture fields stretch south along the lakeside, and the valley climbs behind us in a series of step-like plateaus.

For dinner, we rejoin the family for a meal of rice, barramundi and garden-picked asparagus in a cheese sauce. It really was WWOOFing in style at the DeClare's, and the after dinner discourse was always enlightening. We watch *Iraq for Sale*, a damning indictment of the war, and Anna, who has not had the opportunity to see much in the way alternative journalism before, is thoroughly sickened by its contents.

10/November

The hothouse is still a work in progress, but there's a diversion at midmorning. Gemima is taking their tractor down to help one of their neighbours clear a field, and we decide to join in the fun. Leo and Morag are a delightfully friendly and laid-back couple and they invite us into their house for tea before the work starts. Morag is a midwife and therapist, and Leo a man of many talents whose main vocation is working the land. They are also followers of the self-sufficiency ethic and their house is a mud and hay-bale work in progress.

The tractor makes short work of what would have otherwise been a herculean task, and afterwards everyone helps to clear the loosened rock and boulders by hand. Outdoor work really whets the appetite and Leo prepares a wonderful fresh lunch of pasta and olives for us when we finish. It's the sort of exchange that sums up the spirit of WWOOFing.

In the afternoon we have some free time which we use to explore around the lakeside. Later, Gemima takes us by car up to their other farm property, situated higher in the valley. We get to meet their horses and goats and the rest of their cattle herd. The view from the higher pasture land is breathtaking, and the great rock of Cradle Mountain can be seen framed in a hollow between distant hills, the kind of panorama of the earth that our ancestors revered. For me it was suggestive of primal creation; Cradle Mountain a *Lingam* resting in the *Yoni* of the depression of the surrounding plateau.

Dinner today is a real treat; BBQ mutton, originating from their herd of goats, cooked on their outdoor grill along with fresh salad from the garden. The view from the house is particularly epic at this time of day, as the sun's retreating light steals across the valley, casting deep shadows. The spectacular sight of dappled mountainsides and reflecting lake in the warm glow of setting sun. Fresh fruit and cream complete the dinner and John spoils us with a special dessert wine, which is even sweeter than the fruit.

11/November

Today begins with milking our neighbours. A task which turns out to be much easier second time around. Gemima has brought up some coffee in mugs to treat us to some "cowpuccino" of fresh cow's milk and coffee. Possibly the original inspiration for cappuccino, it is sweeter, frothier and more vital than the usual kind. We herd the cows into their next field before breakfast at the house, and then attend to the rabbits before continuing on with work in the hothouse, which is really coming along.

At midday it's time for lunch at the community hall. Two other WWOOFers from the house up the road, our next WWOOFing destination, are there; a young couple from Brissie; and we set up the lunch. A few local people turn up, a mixed bunch, and the place has the atmosphere of a Sunday church gathering. There is also a co-operative stall out the back and Anna and I stock up on some organic fair trade goodies for our little house.

It is a beautiful spring day and we have some spare time after the lunch. The family suggest a range of leisure activities for us, such as taking boats out on the lake, or hiking up the valley.

We find something better to do.

Later in the afternoon, we rejoin work at Leo and Morag's proto-vegetable patch. Once the work is done, Leo shows us his biodiesel set up; not much more than a few containers and pumps and some old washing machine parts. He explains that it's a small scale co-operative enterprise between four families powering six cars, and that it is pretty much optimised at this scale. The fuel cost worked out at 43 cents (AUD) per litre. I get pretty excited about the cyclical non-fossil fuel element, but Leo was quick to point out that it was not "The Answer". Vegetable oils require a lot of fossil fuel for farm machinery and transportation, occupy ecological space, use fertilizer, etc. ...that all said the small scale setup they had was perfect for getting use out of a waste product. It was another point of note: no one is more informed about sustainability than the people trying to put it into being.

In the evening there is dinner and a film show in the community hall. Most of the small community drift away after dinner and it ends up being just us, our current hosts, our future hosts and their two WWOOFers. A French film noir was tonight's choice, and I was expecting something horribly pretentious. Instead, we were treated to a charming window on a world of the past.

12/November

Another day of hothouse work. It's physically strenuous but extremely satisfying. Unlike so much of what goes on in an office environment, it's challenging, and the fruits of our labours immediate and tangible. So much of what we've seen and experienced in the valley has set my imagination alight, and I brood on the future as I work. Sharing my thoughts during our break times is a release, and we become closer throughout the day.

I'd been feeling more inspired and energetic since I came to Lorien and this feeling is underlined during my afternoon practice. Training stances in the field outside the WWOOF house, looking out at the lake in front with mountains behind; the *feng shui* is just ideal. After the pain of stances, the *qi* flow is exquisite, accompanied by a deep sense of relaxation and peace. The air resolves into stars again, and I can feel the energy pushing through blockages in my system. It's a stunning session, in a stunning place, under a stunning blue sky. Afterwards, I feel tremendously relaxed, calm and happy. Anna notices immediately when I come in from practice; she says I look healthy and that my eyes are bright.

I tell her how fabulous my practice session has been.

We arrive early at the house pre-dinner, and Gemima takes the time to show us the secrets of making a curry base sauce, south Indian style. The curry itself turns out to be delicious. We meet their daughter over dinner, a quiet and pretty girl who attends school in town. Both John and Gemima are noticeably brighter with her arrival, quite the proud parents.

13/November

Gemima, ever active, is off to a meeting in town regarding proposed mine workings in the valley. Gold deposits had been discovered by an Indian company in the great pillars of rock that formed the natural border where the dam was situated. The environmental impact of any future mine workings was a matter of deep concern to the community, and Gemima was dynamic and experienced enough to be a formidable opponent to any corporate or governmental entity that threatened the sanctity of the valley.

For us, there's a fun diversion first thing as we discover that the leader of the rabbit pack has made a break for freedom during the night. A clever boy, even choice carrots can't tempt him back into captivity, and eventually John advises us to let him fend for himself. The rest of the morning is spent completing our work in the hothouse. It is virtually unrecognisable from the jungle of a few days previously, now raked beds of juicy compost, ripe for use. Extremely satisfying.

Anna makes a Korean stir fry for lunch and John joins us. Afterwards, we organise the jar collection in the kitchen pantry. It doesn't take so long to complete our tasks for the afternoon, so we have another fun time of frantic giant white rabbit pursuit, this time ending in success.

Today's dinner will be our last evening meal with the DeClare's. After a week with the family we are now finding our rhythm with life on this property, and it's wonderful to talk and eat with our hosts. We are both touched when they let us know we are welcome back anytime. After dinner, we spend the evening in pleasant discussion, which folds naturally by the commonality of sex; Anna and Gemima talking in the kitchen, myself questioning and listening intently to John, who is a walking goldmine of information, bright and enthusiastic on the wealth of topics that interest them. Anna really appreciated that Gemima was thoughtful enough to speak clearly and slowly for her, as a non-native speaker, as well as all of the teaching and guidance she'd shared with us so freely since we had been there. I enjoyed every conversation I had with John, with his passion and wide understandings founded on a deep sense of human decency.

Today has been a good day. We take our time enjoying the walk back to the house, now used to the dark and the creatures, and bask under the night's magnificent starfield.

Lorien — Part II: Kensho

14/November

There is no rush to proceedings in the morning and we have cappuccino cow milking time, breakfast at the main house, then a tidy up at the WWOOF house. Gemima gives us a lift up to "Kensho", the name of the land trust centred around Charley and Alice's place. We drive a short distance up the valley, down a small private road past a patchwork of little vegetable fields and into vibrant eucalypt forest. The road ends in a cleared space, and here we are; a house and out-house perched at the edge of a steep hillside, looking out between the trees across the valley below.

The main house itself is a delightfully rambunctious little dwelling, which looked liked it had organically grown up over the years by hand ...which of course it had. Despite its small size, it is absolutely jammed full of people; Gemima joins a community meeting around the dinner table with Charley, Alice and a range of *Kensho* residents. Myself and Anna join their WWOOFers in the lounge area; Bronwen and Luke we've already met, and also Jacob, a tall middle aged man travelling with his young son. Our hosts are busy so the rest of us spend that first day mostly on a very large game of monopoly, which Anna manages to win despite never having played the game before. Eventually, the meeting ends and Charley shows us around; our sleeping area upstairs in the main house, the other bedrooms and living space in the outhouse across the yard, the shower block, the small garden that overlooks the steep hillside, and the eccentric touch of the self-composting toilet outside in an old hollowed out tree in the front yard. Charley enthuses over their setup, proudly letting us know that there is "plenty of hot water" and "plenty of electricity" from their micro-hydro generator and grid of solar installations.

WWOOFing proper begins at dinner time and we are all enlisted on the task of making and baking the meal. Somehow I end up preparing the dessert of rhubarb cakes. I misread the ingredients a little (lot) but amazingly it still turns out edible. We get to know our hosts over dinner. Charley and Alice are long time residents of the valley, and first purchased property here way back in 1975. Although similarly "Green" in outlook and lifestyle to the DeClare's, both families had taken very different paths to Lorien. Whereas John and Gemima were professionally successful members of the Melbourne cognoscenti who had decided to completely change their lifestyle, Charley and Alice were self-confessed "hippies"; original members of the alternative movements of the late 60's and 70's.

In the evening we just chill out in the house. Alice was struggling with a foot injury she'd sustained the previous day while out on a walk and Jacob, a certified traditional Chinese medicine (TCM) physician and acupuncturist, performs moxibustion and acupuncture on her. We spend the time playing cards with Bronwen and Luke. He a student and she a post-student, they had driven all the way from Brisbane and were taking a seven week tour of Tasmania. Native Queenslanders, they have decided to stay longer in Lorien in the hope of seeing the predicted snow.

Snow in an Australian spring quite the novelty, but that's Tasmania for you.

15/November

Wake up late-ish. The mellow air of *Kensho* infectious and nothing much is happening first thing. Mid-morning I see Jacob, our tall fellow WWOOFer, practicing some *qigong* outside and we have a good chat about our common interests. Jacob explains that many students drop out of TCM because they simply can't get the change of paradigm; modern western medicine is reductionist, starting from a problem and working downwards to find the details of the problem whereas TCM starts from a problem and builds outward to find its context within the whole system. The science it's most similar to is ecology; studying interrelatedness and the effects of parts on wholes.

Our fellow WWOOFers leave during the course of the morning, Bronwen and Luke disappointed in the lack of precipitation but hoping to find snow on their way up to Cradle Mountain. Alice is doing much better today, and the efficacy of TCM is something of a revelation to her. Alice surprises us with our first job for the day; making a collage of tiles to be incorporated into their patio. Like many of the jobs at Charley and Alice's — right from the start with the cake baking — it threw me off at first. I felt that I was supposed to be working.

But this wasn't working. This was ...more like playing.

Art and crafts are not for me, I thought. But Alice brought the collection of broken up coloured rocks and the basing tray and the cement and spread them across the kitchen table and we got on with it. Anna had no such reservations and was sorting and placing away quickly. *Girl's stuff*. Alice encouraged me to use some flair in the decorations of the border, and showed me how some innocuous rubble could be assembled into the shape of a lizard. The gecko started to form in my mind and I was off and running; sorting through the bags of stones, looking for *just the right one*, and arranging them with more fervour than either of the ladies. The time started to fly and by the time we'd finished the piece I was elated with accomplishment and slightly disappointed that was all there was to do.

It was work ...but it wasn't work. It was great.

The afternoon's task is the clearing out of another house in the *Kensho* trust further up the valley. A very nice property, it had recently been vacated by one family, with another family from mainland Australia already waiting in the wings to take it over by buying into the land trust. The views from this house were stunning. I could definitely see the attraction of both it and life in the valley.

Charley has been away since the early hours, out on a job fitting solar panels to some new build tourist chalets on the west coast. The family run a small renewable energy business from home and it's great to hear that there is a demand for their services. In the evening it's just us and Alice, and we get to know each other better over a card game at the dinner table accompanied by copious quantities of tea. It's the first of many such sessions.

16/November

Our morning task for today is weeding of one of the families' fields. We take the steps down the steep hillside from the house passed the chook's residence, and beyond a woodland path winds down the hill through thick fern underbrush. The path passes over the watercourse that was dammed to feed the micro-hydro plant, the pipes issuing a low hum. At the bottom of the hill the tree cover opens out suddenly, unveiling a large cleared field. I hadn't realised before how big their property was; a land trust shared by several families with the twin aims of preserving the natural woodland and farming the land sustainably; the beautiful old woodland was dense and lush and stretched for acres.

Anna and myself are weeding and hoeing experts by now and it's similarly satisfying work to last week's at the DeClare's. Fresh air, warm sun and plenty to do. We have lunch at the house and are joined by the neighbouring family, who are also part of the *Kensho* land trust. We're finding out that life at *Kensho* is very relaxed, and people seem to pop in and out all the time. Lon and Ciara are similarly chilled out people to Charley and Alice, in this case academics who had decided they wanted to live a life close to the land. Their time in Lorien had already been blessed by a beautiful three year old girl, and Ciara was already showing the signs of another on the way.

After lunch we take the short walk along the private road to Lon and Ciara's, and help prepare their tomato patches; weeding then planting. Meeting up with new people; helping out; having fun; it's the spirit of WWOOFing yet again. In the evening we relax and chat over tea and card games. Card games have become one of my favourite hobbies since travelling, but I'm hard pressed versus Alice, who is quite a shark, card games being one of the social staples of *Kensho* as well.

17/November

Charley is back. We find him at the breakfast table, and he's already been up for hours; milking the goats and preparing the cheese. We catch up over the cheese and milk along with porridge served with copious amounts of honey. All organic produce, the honey coming from one of their neighbour's hives, and all delicious.

Charley asks if I want to accompany him up the valley to install a solar array in a neighbouring property. He should have had no worries there; I almost jump out my skin at the chance. My first solar energy installation. For me, it's as beautifully significant as the loss of virginity. The house is located off another one of the valley's roads, but distances being short in Lorien it doesn't take long to get there. The neighbour isn't a member of the *Kensho* land trust, but the renewability enthusiasm is obviously catching.

Thankfully, most of the hard work had been done previously by Charley; the large plate of sun-tracking photovoltaic panels already hoisted up and attached to its mast. Today it's just a case of wiring it up, linking it to the house's circuits and turning it on. Charley shows me how to screw in some wires and circuit boards, and I get on with it while he deals with the more electrically dangerous details. I get to know him better as the day progresses. Charley, in a relaxed remote rural valley full of mellow people, was perhaps the mellowest. New metaphors would have to be coined to describe how laid back he was. Genial and easygoing, with a dry sense of humour.

He curses under his breath occasionally as he works; he's making a lot of mistakes today, brain fried after the late night drive to get back home to the valley. He's not the type to get angry, but his wry wit is tuned to the task of self-depreciation this morn. After several false starts the array powers up. A beautiful sight; witnessed by just me and the owner's dog; Charley doing the connecting under the floorboards of the house. I'm bursting with pride with my first contribution, however small, to the cause of solar power.

A good job, well done. We head back to the house to be greeted by a wonderful reward. Anna has spent the morning preparing a Korean-style lunch for the family. She has to use her ingenuity as she's pushed for ingredients, but comes up with a lovely rice dish followed by pumpkin pancakes. The pancakes a Korean novelty, even for me. It's great, and it's great to spend time playing cards with Charley and Alice afterwards.

Both Alice and Charley were greying gracefully; happily settled in middle age, surrounded by their large family and their large woodlands in a magical corner of the world that they had discovered and built together. There was a timeless quality to their looks. I could just imagine the two of them farming the land in any northern European country, in almost any time period; Alice matriarchal; dynamic and organised, with flowing hair and a floral sense of fashion; Charley, the tall yet gentle outdoorsman; out under the elements, protecting his charges, or out tinkering in his yard, happily ensconced in a favourite set of well-worn working clothes. I simply couldn't picture the pair of them living in a city; they had a real connection to the earth, country people through and through.

And connecting to the earth was one of the main jobs, nay joys, at *Kensho*. After tea and cards Charley heads off to bed for some well deserved rest, and we head down the woodland path to the cleared field and it's patchwork of vegetable plots. Alice oversees, but we don't need much in the way of direction, and we easily settle into the preparation of a new patch.

18/November

An early rise this morning. One of Charley and Alice's daughter's — most of their kids have outgrown the nest — is home from Uni. She's bringing a friend to WWOOF as well. We greet them at the breakfast table, and it's one of those awkward aspects of WWOOFing; being right in other people's intimate space, feeling a little like interlopers in private family affairs.

But they are very friendly and accepting of these two strangers in the family house. Davinia a good looking young girl who is positively two chips off the old blocks — you could really see the similarities with both her parents — and her friend Nastassja a lovely young girl of Sri Lankan heritage. Tea and cards is the perfect way to break the ice.

Today is a slow Saturday and not much is happening. We do some work on the vegetable patches in the morning before attending the hall for the official Saturday lunch. Me and Anna again take the opportunity to buy some chocolate from the Co-op and then relax outside under the sun, appreciating the weather and the view.

I don't want to give the impression in my writing that Lorien was some sort of 'Green Paradise'; it was not: it was just real life. Real life in a remote highland rural community in Tasmania, which is much like the real life that goes on in remote highland rural communities everywhere. The valley was an eclectic collection of families and individuals, all with their own ideas and outlooks, likes and dislikes, friendships and grievances. And not all of the people were 'Greens', there were those considered 'Rednecks' too, including some that owned the hated plantations. But it was a beautiful place, and what Lorien did have was a lovely sense of community.

After some more tinkering in the vegetable patches, we help prepare some dishes for the Saturday Dinner at the community hall. The dinner tonight is quite something, with a host of community members preparing some lovely dishes, including some excellent as usual preparations from Gemima and more fine wine from John, and we have a greedy time sampling all of them. For movie time tonight, it's *Dreams* by Akira Kurosawa. It's a film that stirs some memories of the distant past for me. A wonderful work, its idealistic theme was so unbelievably fitting of our time in the valley. Alice sums it up perfectly at the end:

"Beautiful!"

On the drive back up to the house, another unexpected treat; Charley jams on the breaks as a Tasmanian Devil runs across the road! Everyone's excited; they must be a rare sight even for the locals. Caught in the headlights, it waddles hurriedly across the road and up into the trees. It's larger than I expected, more like a marsupial badger than a little dog.

What an amazing souvenir.

19/November

Another relaxing day at *Kensho*. Rather than starting work straight after breakfast, we instead have a big game of cards; now more advanced games are the order of the day with such a crowd around the table.

Take our time to get down to the fields, and it's a case of many hands make light work with Davinia and Nastassja joining in as well. It was a delight to be out in the fresh air, under the southern sun, working away. The earth was black and rich and full of life, with living new discoveries under every trowel full of dirt. All manner of bugs and grubs, many with that rather terrifying prehistoric Australasian appearance, befitting of titles such as 'witchety'. We took occasional breaks, to rest and feel the warm sun, or appreciate the wind blowing in the green trees that surrounded us. It was beautiful. Looking up at the cloudless blue sky, with hardly an aeroplane ever in sight, it felt remote, yet also homely. Sanctuary. It was easy to forget about the outside world in this place.

There was a lovely rhythm to life at *Kensho*. The work in the fields or with the animals was always enjoyable, along with the camaraderie that went with the shared outdoor experience, and we equally looked forward to break times back at the house, whether lunch or dinner, tea or snacks, and the fun and games that went with it. The house itself was such a cosy dwelling, full of life; a warm front room of comfy seats, walls brimming bookcases; a wide bay window providing a fantastic view over the woods and also serving to illuminating their study area; the kitchen upon a dais at the far end of the house. The kitchen was centred around a big old wooden table, from which pulsed the heartbeat of the home; all discussions and decisions in the household seemingly radiated out from meetings at this distinguished old desk, alongside endless cups of team and endless games of cards.

We had finished the planting in the field, so after yet more competitive rounds of cards, the girls had some household tasks set, while I went up the valley with Charley in his truck, to assist with the removal of some old deconstructed pig sheds from the recently vacated property. There was a lot of heavy lifting to do, but Charley was a great guy to be around. He didn't even laugh when I threw a sheet of corrugated iron away in panic as a disturbed Hunstman — the biggest non-tarantula spider you've ever seen — scuttled at light speed towards me.

20/November

We completed the metal moving in the morning, and it took several shuttles up and down the valley to transport all of the scrap to their front yard. I had a feeling that it wouldn't go to waste. Their front yard and the retaining wall lining the private road were a noticeably massive set of earthworks, carved out of the hillside and lined by huge rocks. I was surprised to find that they had been built by hand and not machine; WWOOFer powered construction! The family may have been laid back people, but the effectiveness of their own work ethic was highly impressive.

The other days' pancakes must have gone down well because Anna had prepared a brilliant vegetarian sushi lunch. *Kimbap* as the Koreans say, and I make a pig of myself after all the exertions of the day. When we're at the table, I take as many opportunities as I can to quiz our hosts over everything; their business, their renewable energy systems, their farming, their philosophies, their backgrounds; and after lunch Alice shows me some press clippings from a time when their treasured home and valley were under threat.

The "Battle for Lorien" was fought against the forestry industry and the authorities; the family had been settled happily for years at *Kensho*, their private corner of the world, the trust they had co-founded to protect the woods and landscape of this beautiful Tasmanian valley. But while *Kensho* was entrusted and protected, and the community of Lorien shared in their ethic of conservation and sustainability, the forest on the high plateau above the valley of Lorien was not.

Forestry as a business moved in; clear cutting the original forest, destroying the native ecosystem, planting monocultures of trees, poisoning the wildlife and spraying their new crops with herbicides and pesticides. The Lorien residents revolted at this abuse of their water-catchment area. Protests were organised and a political movement formed. But the forestry business could not be moved, and considered all of their activities acceptable, and all of their husbandry legal and within national guidelines.

With industry and government impassive, the family had to take matters into their own hands. They collected samples from the stream that supplies their house and lands with water and sent them to an independent laboratory in Melbourne for analysis. The results came back. Not only had the herbicides and pesticides penetrated into their water supply, they had done so at levels several times higher than stipulated World Health Organisations safety limits. With this new ammunition, the family returned to the fight, and the community was vindicated in their stance against the plantations.

Lorien won the battle.

With our work finished for the day, we laze about in the house before preparing pasties for dinner. Another mammoth card war crowns our evening.

21/November

Our final full day at *Kensho*. We were back up at the vacant property, continuing with the business of getting the house and grounds in a fit state for the new arrivals. The past residents had thoughtlessly abandoned an old car at the far end of the garden. An old 70's Ford, now ruinous, sitting in a hollow and completely covered in ivy. Alice was fuming that she was left to dispose of it, and was seriously considering legal action to cover the costs of the removal and scrapping.

1970's Australian cars were big machines, and this *Mad Max*-extra was going to take some work getting out. Everyone piled in, clearing the weeds out of the way to reveal a bright orange road monster, its paint job still in surprisingly good nick underneath all the weeds. No wheels or engine though; it was going to have to be dragged out of its premature resting place.

Alice hadn't actually known how on earth they were going to be able to dispose of it, until Superman came to the rescue and offered the use of his trailer to wheel it to the scrapyard. Superman being the *nom de guerre* of one of the Lorien residents, known as such because of his habit of painting the Superman symbol on both his cars and his hives. The local beekeeper, he made voluminous and excellent honey. He had just bought a new multi-purpose trailer and was graciously living up to his nickname by lending it out to his neighbours in need.

It took a full morning to clear the car and a path for its removal. Charley hooked the orange hulk up to his truck and pulled it with no little strain on his engine out of the hollow, out of the garden, and up onto the road. Another helpful neighbour was on hand with a small tractor to crane the beast onto Superman's trailer. The dinky little trailer heaved as the strain of the giant motorcar was gradually placed upon it. Eventually, the weight was fully released and the wires disconnected...

...a moment of silence from the crowd...

...and then the little trailer buckled down the middle.

"Ohhhh, my bloody trailer!!!"

Superman was not happy. His brand new little all purpose pick-up had just been destroyed on its very first mission! I felt for him, and was a little embarrassed for our hosts. I hoped they had insurance.

Davinia's boyfriend Ralph arrives in the late afternoon, his arrival christened by yet more tea and cards. Again, as with the DeClare's, I feel that we are beginning to find our groove just as it becomes time to leave. Our time here had been a growing marvel; such a safe, happy and exuberant place. Life at *Kensho* was like a living web; centred on the dinner table of the family and the fun and games that accompanied the communication that was the lifeblood of the home. There was always something to do, but never any great rush to do it. Life here had such flow. Visitors and family members came and went as they pleased, and something new or interesting was always around the corner. The family didn't seem to want for anything; whether time or money, food or drink; there was real abundance. There was no better way to describe it than with Alice's choice phrase:

"Beautiful!"

The family let us know we can stay for longer if we wish, another touching sentiment for us here in Lorien, but we regretfully let them know we already have a planned schedule. Cradle Mountain is calling.

22/November

We say our goodbyes. Crossing the threshold for the last time we step across the little collage we made together. It's both wonderful and poignant to reflect that we have made our mark, however small, on this little valley on the other side of the earth and equally that this little valley has made its mark on us.

Ralph works as a park ranger, and he and Davinia graciously give us a lift to cradle mountain. As we pass beyond the hydro dam it feels like crossing a limin. We're no longer in Lorien. Driving along the surfaced and evergreen-lined highway, we could be anywhere in the western world. It begins to rain. Clouds fill the sky and the wind increases, driving the rain against the wind-screen. I think wryly; now it doesn't feel like just anywhere in the western world; now it feels like Scotland. The weather continues to deteriorate as we head into the park. Davinia and Ralph drop us off at our accommodation and we thank them for their generosity.

It's blowing a gale but we decide to get second breakfast at the park centre and then brave the conditions on the trip to the eponymous mountain. We walk the first part of the wildlife boardwalk and almost die. Despite jumpers and raincoats we're not well enough equipped and we start to get very cold very quickly. It's an interesting journey nonetheless, through strange alien scenery. Moor-like but more forested than the British equivalent. The flora is very different. Again that Australasian twist; some of it wouldn't look out of place in Alice's wonderland. Strange tufty mounds of greenish-brown grasses populate the landscape, like great flocks of inanimate sheep. Or mome raths. The exotic scenery is composed of a multitude of wet green and olive hues, and forest-wrapped rocky outcrops emerge in the distance occasionally from behind the billowing white clouds that are low to the earth in the storm, making for truly atmospheric scenes. But it's definitely the sort of atmosphere that is better to see on TV than to actually walk about in.

We are rescued from our ill-prepared hike by one of the parkland buses, which drops us off by Dove Lake at the foot of Cradle Mountain. It is bitterly, bitterly cold and we have to huddle together for warmth. The swirling foggy clouds have reduced visibility down to 100 metres or so and the iconic Mount is completely occluded. We take a quick look about the lakeside but the driving wind and icy rain defeat us, and we return to the cabins to shower and recover.

23/November

The conditions haven't improved, and we decide to head on to Strahan. The weather clears as we travel away from the park, and the drive through the moorlands, forests and mountains of the west of Tasmania is truly beautiful. I noticed I'd been using the term beautiful a lot recently, a habit I'd picked up from Alice. But it was fitting. This island really was beautiful.

24/November

A day of exploring Strahan. A tiny colonial-era town on an estuarine river. A sleepy place, whose past identity was founded on the forests and minerals of the west coast. Tourism was the town's modern specialty, an ideal choice considering the beauty of the region. The town specialised in carved trinkets made of Huon Pine; endemic ancient trees that could live for thousands of years, the majority of which were clear cut long ago. The fragrant wood was now sourced from the stumps the logging industry had left behind. We enjoyed our time and decided to stay on another day.

25/November

Ocean Beach. A 15 km or so walk to the beach along a very quiet small track across gently undulating dune lands. The beach is spectacular; stretching to either horizon, mile upon mile of white sand backed by high steep dunes. Waves of bright surf break and roll across the shallow sands, their understated majesty a hint of the quiescent power of the Southern Ocean. We have our packed lunch sitting on the dunes, appreciating the quiet of remoteness. The immensity of the elemental view brings to mind the temporariness of not just our time here together, but of our lives; intensifying the moment. It is a place to think about who we are. We didn't know where we were going.

26/November

Strahan to Hobart. We change buses in Queenstowne, the inland mining town that originally fed the port of Strahan. The cleanliness and quaintness of the buildings in the town, nestled between picturesque woodlands and water-courses, stands in utter contrast to the surrounding hills. Heaps of mine refuse surround and abut the town and the peaks of the local mountains have been *removed*. Completely destroyed. As if a giant with a pair of shears had cut off and thrown away their conical peaks. It's bizarre. I also can't understand how it could possibly be healthy to live right in the middle of a devastated wasteland like this, surrounded by mine tailings and the dust of disintegrated peaks.

I'm surprised as the bus route to Hobart climbs directly up and over one of the degraded mountains. If the view from the town was bizarre, it can't quite compare to the view from the top. The road flattens out as we travel across the planed peak. I realize we are not traversing over a massif. We are traversing over a carcass. It no longer looks like a mountain; the innards have been exposed, a mass of bright and gaudy colours marble the exposed bare rock, the hitherto occult organs of the mountain, and then trail down the hillsides; exposed entrails. All of the peaks surrounding the town have been ravaged in this manner. I'd never seen anything like it. The death of mountains.

A shocking sight to reflect on as the bus wheels onwards. The view for the next few hours provides more comfort; Lake St Clair National Park a stunning rolling panorama of peaks and forests, lakes and valleys. The sheer size of the protected areas in Tasmania is heart warming, and a wonder to behold. Eventually we reach a pastoral upland area in the centre of the island which illuminates the extreme contrasts of the island's climate zones. We cross a high hill; west, the view is lush green and mountainous, east, the low plains are dry, golden yellow and denuded of vegetation. This pastoral upland the perfect median; hilly, with broken tree cover and olive hues.

The road continues along the course of the rising Derwent river; the rolling view now a pretty unfolding scene of quietly building civilisation as we move closer and closer to Hobart.

27/November

A day of recuperation after recent travel. The only stress being fear of pregnancy.

28/November

We decide to pursue all of the tourist activities we can afford before it's time to leave. It turns out to be the perfect day to visit the peak, with good visibility under a bright blue sky. It's remarkable to experience such a high peak so close to a town and the views are exceptional; a 360 degree panorama of the south of the island. The town below girdles the channel, the bay a deep sky mirroring blue; the Derwent valley stretching north; hills in the golden east and rugged mountains in the distance in the west. The arc of the southern horizon is deep blue Ocean. Awe-inspiring.

29/November

I love chocolate. So I'm a big kid again as we visit the local factory. The cool climate and waters of Tasmania were perfect for milk production and thus the generation of fine dairy chocolate. It's another chance to have fun with my girlfriend and treat her like I would like to treat her at home. Just lovely. Back at the hostel we spend time talking about how our lives have changed since we've been in Australia and the book she was currently reading, *The Alchemist.* I had read it before I travelled. Such a beautiful book. And so true. It was one of the reasons that I decided to follow my heart and experience the world. Reading it again I could see with a smile that I was the boy in the fable; I had been on my journey, and I had discovered my treasure.

30/November

Melbourne to Hobart. Another excellent day for a flight. I felt close to this island now. The experience had been more than I ever could have dreamed. Tasmanian Tigers waved us goodbye from the airport gift shop. Visitors are soon to discover that Thylacines are omnipresent in Tasmania ...except in actual existence. It could have been a sour note to leave on, but I no longer felt like lamenting the plight of the beast; the past is dead. It cannot be changed.

I was looking to the future now; full of optimism after our time in Tasmania. Experiencing the care and attention the people of Lorien put into their valley; of how people could enrich the land as the land enriched people, I was only encouraged. I could only think about how beautiful Tasmania was, of how the hellish conditions of the early penal colonies, of institutions like Port Arthur Jail and the Female Factories had been transformed over only a few generations into the safe, peaceful, bountiful and happy place that we had discovered. Moreover, the WWOOFing experience had felt like riding the crest of a wave; the crest of a wave on a sea change that was taking place throughout society and throughout the world.

Both eyes were open.

Leaving

02/December

I had spent the past two days rushing around trying to get everything in order before I leave. Just appreciating every last morsel of time with Anna. Our minds were both firmly fixed in the present.

It was a delight to be back amongst familiar faces. Kelly had welcomed us back from the airport and Anna was more than happy to switch back into Korean mode. Kelly and three of the other girls had moved from the hostel into a fantastic city pad, and they generously put us up in their living room and threw a little party for our 'homecoming'.

Bastian had moved onwards and upwards from hostel life as well; he had got himself a job with a software company, and had even managed to bag himself a yuppie flat near the seaside through a house-sitting deal! He was a hard man to stop.

New Zealand

03/December

It was a strange goodbye. We just hugged and said the words. We didn't promise each other anything. We didn't know if we would be able to meet again.

I felt numb at first. I stared out of the window at the ocean below. The plane was pulling away rapidly from Australia just as I was separating from Anna. I thought of our two lives, so small amid the vastness, as I looked out at the great mass of the earth. Like two little stars within the infinite void of space; once blended, now pulling apart in every way possible; physically, mentally and emotionally. A separation paralleled by the inexorable movement of the jet at speed away from the land.

I was emotional. I tried to distract myself with the in-flight entertainment. I was faced with the movie *Superman Returns*. I didn't think it was possible to feel any worse than I felt already but I was wrong; I now had to put up with this product of the demented Judaeo-Christian psyche. What is everyone so afraid of that they need a Messiah to protect them? What is Superman so afraid of that he needs such big muscles?

The film reminded me of 9/11. Where was Superman when you needed him?

There was no Superman and there are no Messiahs.

There is only us. We are responsible. The message of the great spiritual teachers throughout the ages; we are responsible. Not Jesus, not God, not the Buddha, not Allah, not Krishna, not Superman. Me and you.

Amongst the cries of disbelief on that fateful day in 2001, and the vengeful gearing up for war in the months and years that followed, something important was forgotten. The official causality of the act of terror was a circle. A circle which led straight back to the Pentagon, and the training and arming of Osama bin Laden and the Taliban.

Agents of terror for the ends of the U.S.A.

As you sow, so do you reap.

And beyond my reverie an even larger sense of disbelief. I sat and looked around at my fellow passengers. What did they know of themselves? Every one of them, and everyone alive, has what could be termed supernormal abilities.

But they do not know it because they do not believe it, and they do not train the mind.

I see New Zealand before I see New Zealand. A sheet of dirty black smog fills the distant horizon before the island itself is visible. Perhaps not the best herald of a new land, but my outlook brightens as the landscape comes further into view. It is something of a revelation; after the desiccation of Australia the vibrant greens of the North Island are a veritable feast for the eyes; a bountiful landscape of hills, forest and pasture; the coastline a fractal curve of bays and sandy inlets spreading into the distance, set against aquamarine seas. As the plane banks into its final approach, the modern city spreading around the emerald cones of slumbering volcanoes comes into view. The remains of archaic hill forts stand atop many of the peaks. It reminds me of home, but the greens are brighter and the vegetation more lush. It's electric. After Australia from the air the sight is just extraordinary. I can't wait to get down and explore.

I was greeted at the hostel reception by the biggest Polynesian you've ever seen. A massive man with a ready grin and the unexpected attribute of one of the most quintessential cut glass English accents you've ever heard:

"Excuse me while I put on a video for the children"

He said politely, as he excused himself to put on a DVD for the hostel crowd during the administration process of my arrival. He seemed to take great pleasure in the juxtaposition of his stature and his eloquence and was one of the most convivial and effortlessly humorous hosts I'd encountered in my whole world adventure.

It seemed like a nice way to arrive.

04/December

Exploration. I wandered in the general direction of the city museums and walked into a Bollywood shoot that was going on in one of the many beautiful public parks that populate the city of Auckland. Such a curiosity, and I kept my eyes peeled for any of the female superstars that inhabit such motion pictures.

Aishwara Rai was keeping a low profile so I headed on through the bright new world streets to those volcanic cones and bright green fields I had seen from the air. They turned out to be a fantastic place for packed lunch and the recharge of practice. I noticed, wryly this time, I was once more within *the Domain*.

The museums of the city were beautiful and fascinating places, full of knowledge about the spectacular and unique flora and fauna of the isles, and of the dynamic geology. One of the exhibits that sea level rise resulted in tectonic stress and thus increased incidence of earthquakes and vulcanism. Yet more fuel for the global fire, as if any were needed. I spent the most time discovering the culture of the natives, and the story of their colonisation, that same sad story repeated throughout the Europised world. The particular tale on the walls of the national museum ends with what the writers must have considered a message of integration and hope for the future. But to me if felt like a sad capitulation to assimilation, coming as it did at the tail end of a history of conquest and domination; exploitation and serial betrayal.

05/December

I'd heard good things about Rotorua, so I caught the early bus. The ride comfortable and not too busy. The suburbs of Auckland stretch for miles. Eventually, the scenery becomes more rural; very pretty, green and lush. The journey takes us through a region of meadows, woods and fields. Miles upon miles of rolling hills and wooded pastureland passes by and a spine of low dark hills forms a backdrop in the north. The distinctly Tolkienesque look is crowned by a signpost we pass giving directions to "Hobbiton". This part of the world really was a good choice.

The hostel in Rotorua is extremely quiet but nice and clean. I've only one roommate in the dorm and we strike up a conversation. Rod seems a very nice guy, and is one of those rare creatures: a traveller from the United States. He gives me some good advice on things to do and see in Rotorua.

The town of Rotorua is set right on a great lake, and I wander down to look around. The colours of the landscape are vibrant in the Southern Hemisphere light, and green pasturelands and blue sky contrast with the milky white volcanic lake. I'd never seen real vulcanism before and the sight and sound of the scattered patches of steam rising and bubbling cracks in the ground was enthralling. The smell not quite so enthralling. The parkland meandered along the lakeside, and was filled with a variety of birdlife including various pretty and unusual New Zealand varieties. An island in the middle of the lake and the hills which rimmed the whole region served to underline the elemental nature of the place; it was all one massive volcano; the circular hills a caldera and the island a new cone; yet strangely this lurking annihilation did not make it any less habitable. Such a remarkable landscape; the quiet little town backing directly onto blasted earth, sulphurous pools and the inky white lake itself.

I catch up with Rod again over dinner and have a good chat. He's a very interesting guy and he waxes lyrical on his reasons for travelling in New Zealand. An engineer with a Masters Degree, he really did not like his first immersion in corporate culture — the egotism, the backbiting — and wanted to do something different before settling down. He had to split up with his girlfriend of two years to travel, but felt it was a necessary step because of their divergent philosophies of life. Despite being a good looking, square jawed, All-American on the outside, he launches into an impassioned critique of American culture; he didn't like where the youth were going; the effects of the lifestyle on their attention span, their lack of education, their poor diet. Parents have no time for their kids and the society seems programmed to go to school, university, get a job, work for your whole life, retire. Then maybe travel or do something different. Working life itself consists of long hours, long commuting time and a diet of fast or convenience food because of the draining effect of the long hours. Each day repeat. He finished:

"The whole capitalist philosophy of *greed* seems to underlie it all."

I was both surprised and overjoyed at this unprompted self criticism. Even inside the Belly of the Beast people were beginning to wake up just like I had.

I advised him to try somewhere completely different before his travel experience was over — you can't really learn anything in your own culture — and I tried to think up a good destination...

"*Cambodia.*"

Rod jumps into the gap in the conversation, in a moment of inner illumination. I had no doubt that Cambodia would be a brilliant choice.

We wash up our dishes in the kitchen and run into a bunch of blonde Scandinavian girls, who I couldn't help but notice when they were occupying the hot tub earlier in the day. The girls were eager to talk and displayed the body language of openness to sexual approach of the young adult on holiday. I have a tendency to freeze when hit by an unexpected overload of beauty but Rod manages to bullshit through a conversation of sorts with the "aw, shucks" outgoing charm that is one of the best advantages of American culture.

Later on in the dorm he extends an invite to me to go out on town from the Norwegian girls who are looking for "some men to talk to for a change". I thought it bad form to jump straight back into the zone of temptation just after saying goodbye to my girlfriend and decided I'd prefer something mellow. Rod was all partied out so an alternative plan was formed of going to one of the volcanic night swimming spots. Other hostel dwellers interested in the excursion were Belle, an English Ph.d. student and Lothar, a big German fellow. I never in my life thought I'd turn down a night out on the town with a bunch of holidaying gorgeous blonde Scandinavians, but there you go.

An unexpected night unfolds. Rod has a car and we drive along woodland roads to the thermal creek. The water of the creek is clear and the steaming heat is obvious in the night air. The water itself is extremely hot — not merely warm like the spring in the Northern Territory — but not hot enough to burn. The water is shallow and the bed sandy and I sink down and float about, Rod and Belle sit down and soak, while Lothar is happy just to sit on rocks and warm his feet. The night sky is beautifully starlit where it can be seen between the branches of the trees that line the creek.

Rod and Belle have an animated discussion chat about Green issues, which I reflect on as I bask in the heat and watch the stars.

"Y'know I've been reading and solar is not the answer; there are toxicity problems associated with manufacture."

"Yes but those problems are being overcome by new ways of manufacturing!"

"But what about the albedo effects of all those panels?"

...and so it continued. On and on. Engineer and scientist. I contrasted their ways of problem solving with those employed by my former hosts. Rather than bounce repeatedly off barriers existing only in their own heads, why not start from what already works on the ground in the real world and build from there? I admired them both; they were extremely bright well educated young individuals, with good hearts. However, their minds were like train tracks; the linearity of their thinking was obvious.

We sat out in the moonlight and cooled down. Even in the cool night air it was perfectly comfortable to stand around due to all the radiating heat from our bodies after the sauna of the river. The cooling down process was exquisite ...and then another dip in the river and some more chat. Lothar came from a military background rather than an academic one and was a little left out of the discussions.

On the drive back we stopped at a very large boiling mud pool. It was quite a beautiful sight in the moonlight, and the moonshine through ever-greens cast bright rays across the steaming crater. Continuing the theme of the day, I ended up having a conversation with Belle about China. I wanted to talk about the effects of seeing the scale of the pollution there myself and how it had shaped my evolving worldview but she cut me off:

"But we can't preach to them ...after all the likes of the U.S. have still to sign Kyoto."

Kyoto. A band aid for a punctured jugular. Whereas I was wanting to go deeper into this largest of issues, she preferred to kill the discussion at a superficial level.

No one wants their base assumptions pulled out from under them.

Once we got back I said goodnight to my new friends, and went to bed more determined than ever to write a book.

06/December

On Rod's recommendation I take a day trip to one of the thermal parks of Rotorua. It's quite a stunning place, with the wonderful contrast of lush New Zealand vegetation and venting rents in the earth. The spectacle of a large geyser sets the tone for the day and I spend my time wandering around the various chemically coloured lakes, bubbling mud pools, smoking craters and assorted earth wonders that populate the park.

In the afternoon meander round to a Maori village tourist attraction. The village itself not too inspiring but the craftwork in the village shop is excellent.

07/December

On the bus back to Auckland I randomly end up sitting beside Lothar. I get to know him better during the journey and he's an interesting man with some thought provoking opinions, though I can't say we are on the same wavelength. I grow to respect him but I cannot say I like him; a tattooed bodybuilder, his attitude contains an unspoken sense of superiority towards me.

Check into the hostel and then do some souvenir shopping. I'd been keeping abreast of developments in Fiji since started getting unstable back in October; now there was a full blown military coup under way. The UK Government advised against travel. But I decide to go. With a 4 a.m. start to catch the flight I have an early night.

Fiji

08/December

I catch a taxi to the airport with a Danish girl from the hostel. Air flight is always a reminder of mortality, and this morning's even more so; a flight into a country where a *coup d'etat* is underway. It's a noticeably smaller aircraft than I'm used to, lacking movie screens and entertainment and it's only about half full, the circumstances undoubtedly being a dissuading influence on travellers.

The flight itself is silky, beginning with placid views of the north island in the pre-sunrise glow, its vibrancy dormant in the night. For most of the way my window lets in a sight of bright sun shining over speckled clouds and ocean far below. Eventually, the clouds on the horizon change in shape and size, the harbinger of land at sea, and we begin the descent to Viti Levu, the main island of the Fiji archipelago. An exotic shore comes into view and the plane drifts over a patchwork landscape of gardens and plantations. Volcanic mounts rise in the distance, scarified with dark green forest.

Three banjo playing Fijians in loud shirts greet us on arrival at the airport. I don't have any concrete plans, so I go with the flow and take the tour guide's recommended beachside hostel.

The dorm is quiet and I make acquaintances with the young English traveller I happen to be sharing it with. Once settled, I take the short walk down to the beach. It's unexpectedly cool and cloudy yet strikingly exotic. Standing on the pale sands of the wide bay, the view is of aquamarine sea under light blue sky and scudding clouds. There are distant hills on the horizon to the north; those volcanic mounts I had seen from the air; a patchwork of greens. A donkey is tethered to a palm tree besides the refuse of a bonfire and the sight sums up the place; a place of contrasts; tropical beauty with the run-down air of poverty.

Looking out at the sands and the sea, feeling the breeze and the tropical uniqueness of the place, I really, really start to miss Anna.

09/November

I take the bus into Nadi to get a feel for the town. The juxtapositions of yesterday apparent once more. Small fields, plantations and market gardens line the roads, the homesteads made from corrugated iron. The bus crosses a bridge and I can see families washing and playing and swimming in the brown river amongst the dense green vegetation; the pattern of ages.

In the town centre the building quality improves as tourist boutiques and multinational franchises, largely run by the Indian community, populate the streets. In the streets themselves there is some squalor; beggars and drug addicts, mostly indigenous Melanesians. Those contrasts again. Natural wealth and material poverty. The overall feel is grubby and post-colonial.

With nothing much to do in the hostel or in the town, and advised against travel eastwards towards the capital in view of the political upheaval, I take the safe option and book a tour of the islands off the coast to the west.

10/November

The cruise through the Mamanucas and Yasawas is enchanting. The Mamanuca Islands take the form of a series of rocky outcrops emerging from blazing blue sea, covered in scrub and fringed with palm trees. The Yasawas are smaller outlying islets which are true coral outcrops; 'desert islands' in the flesh. They are a real sight to behold; stories from childhood and a thousand TV adverts come to life. What look like dragonflies pass in clusters around the wake of the boat. With no little sense of wonderment, I realise these little darting shapes, so agile and rapid through the air, are actually flying fish.

For the first two days of the tour I'll be sampling life on the catamaran *Wanna Taki*. Say hello to the other guests over lunch and then take part in the jumping off-the-boat competition later. I get to know the people gradually as the day goes by; most are like me, backpackers on the closing leg of a grand tour. We go on a snorkelling trip in the afternoon and take in the beauty of this quiet little bay where our catamaran is moored. The underwater life is the richest I've seen; a coral wonderland teeming with bright fish, sea cucumbers, eels and rays.

"*Bula!*"

The native Fijians are an outgoing people and their hearty greeting encapsulates their expressiveness. Despite the tourist industry being just that, the friendliness and warmth of the locals is genuine. We are treated to a serenade by the crew over dinner before being offered the traditional greeting of a toast of *Kava*. The social drug of choice in their culture, it looks like a bowl of muddy dishwater. The taste is not far off muddy dishwater, with a distinctive peppery ginger undertone. Rooty. It has a noticeably numbing effect on the mouth and a pacifying effect on the brain. It makes you very relaxed. Socially lubricating at first, positively drowsy later. It's like drinking mild anaesthetic.

The sun sets over a paradise scene and I reflect that all that is missing are loved ones to share it with.

11/November

A morning diving excursion to the *Garden of Eden*. I join up with two young American guys from the boat and a young local guide takes us out. Bouncing over the waves to the site we can see that the water below alternates between deep and shallow. With no little anticipation we dive down and follow our leader.

We sink down to thirty feet or so, finding ourselves in an undersea grotto; all around the walls of coral cliffs rise and curve up and over our heads. It is marvellous, and I realise that I've seen the sight before; on the pages of a National Geographic in my youth. The corals are brighter and more varied than I've ever seen elsewhere and we spend an incredible morning weaving our way through the towering forests of living rock, marvelling at life under the sea.

The Americans are good guys, two twenty one year olds from the mid west, who had both spent a semester studying in New Zealand. We hang out in the afternoon on the *Wanna Taki*, alternatively chilling over beer and cards, or messing about diving and swimming from the boat. Tyrone and Simon are interesting and bright young guys, full of the dynamism and optimism of youth.

Characters both; Tyrone an easy going young design student who was reading *Zen and the Art of Motorcycle Maintenance*, Simon a medical student who conceded in a beer haze that he funded his extracurricular activities through sperm donation. He gives us a toe curling run down on proceedings, including letting us know that the hardest thing about it was that you had to abstain in the weeks leading up to donation:

"Now I don't get to dick many chicks ...but I sure do like to masturbate."

Apparently he'd been discontinued as a donor at several clinics; meaning that he'd already been responsible for ten progeny with that particular clinics clientele. Myself and Tyrone are also blitzed on a combination of beer and *Kava* and don't spare in flaming him over the long term effects of his fecundity.

12/December

Another peaceful day in the South Pacific. A morning of snorkelling where I'm rewarded with the sight of a reef shark circling in the distance and then lunch and more cards with the guys. Myself and Tyrone are picked up in the mid afternoon for our transfer. We say goodbye to Simon; it had been great to meet him. It had been great to meet both of them and I was disappointed our hanging out time had been so short.

South Sea Island was one of those little coral atoll islets I'd glimpsed earlier in my trip through the *Mamanucas*. I never thought in my life that I'd ever be on one of those little 'desert islands' but here I was. Couldn't be more than three feet above sea level, made of sand, a covering of palm trees in the middle and some cabins for accommodation summed it up. With a diameter of only a few hundred yards it was slightly claustrophobic. There wasn't much to do in the evening besides racing hermit crabs and drinking games and I passed on both.

13/December

A tour through the *Yasawas* on a sailing ship. As holiday booze cruises go it was exceptional; the sights, the sounds, the beaches, the snorkelling, the food, the value for money. But as with everything I did on Fiji, the absence of a certain young lady was felt. The more amazing the experience, the more I wanted to share it with her.

14/December

I had the opportunity to go on my first ever wreck dive today. It was an exceptional morning. As I looked out during my practice on the beach the sea all around was calm and still. I almost couldn't believe my eyes, it was sea; Ocean in fact; but it was as smooth as glass. All the way to the horizon. *There were no waves.* Just gentle ripples on water that was translucently clear. To see something so mighty as the Pacific so still was simply extraordinary.

The boat pick up came in the early evening. Simon was also onboard and we caught up. I was reassured to find out that it hadn't just been me that had been in a reflective mood recently. It felt like the end of the summer holidays; all these crowds of young people getting ready to go their own way after a break time of freedom and fun. It was strange seeing him again; once just a face in a crowd, now a human being of living colour through the time we shared. I thought about the fact he had sired tens of children to my zero, all by age 21. *Oh well,* I thought, a charming, good looking young doctor-to-be isn't the worst choice of father. We said goodbye and wished each other well.

15/December

I checked back in to the beach hostel. There was an eighteen year old French boy there who had spent the last two weeks travelling all over Fiji on the local buses, including into the centre of the capital. I asked him what it was like. He said that everything was fine and life was as usual, except for the complete lack of tourists. I don't know if it was innocence or ballsiness that had fuelled his decision making but it sounded like he'd had a great time.

I hadn't been much worried about the upheaval. In the days leading up to the *coup* the leader, General Bainimarama, had called it off on a Friday because he wanted to watch the rugby match on the Saturday. It was that sort of *coup*. That was not to say that it was without its effects; the people were obviously suffering and that long name, *Bainimarama* was on everyone's lips. From the hostels to the streets to the cafes to the shops, the political crisis was everyone's concern.

It was with some chagrin that I caught up with world events through that incredible medium of the internet. The *Baiju* — the Yangtze River Dolphin — had just been declared functionally extinct. A creature I had shared life on this earth with was no more; not one of them, all of them. If you cut the river in half, you also cut the Dolphin in half. When will we learn?

16/December

Souvenir shopping before it was time to leave. A case of *Bula* shirts all round.

Crossing the Pacific.
The plane is remarkably no frills for such an epic journey and I spend most of the time reading and writing. The view from the window endless similitude. Scattered clouds below and the frozen plastic of Ocean beyond. As a child I'd looked at the map of the world in awe, drawn my little finger across those distances, around the globe of the earth. Now as an adult here I was. I was doing it. Making it real. But as a child I could never have guessed how I'd now feel. A decidedly mixed sentiment; the triumph of the voyage and the guilt over its effects. I drift off to sleep...

Into the Belly of the Beast

Slow descent and Los Angeles becomes visible through the low cloud. L.A. and urban sprawl associate in everyone's mind; however to fly at speed over it and see nothing but an unfolding miles upon miles of low rise buildings is a completely dumbfounding experience. The only patches of green I could see were golf courses, the rivers merely dead concrete channels. One huge, high energy, high consumption urban desert. Planet earth contracting a case of concrete cancer.

I encounter a problem when the hostel I planned to stay at is fully booked out. Go for a wander in an attempt to try and find an alternative and realise I'm on the walk of stars. Wow! As I walk along the power and the glamour of American culture sweeps over me; the music in the shops, the faces in the windows, the names on the street. A thousand treasured memories and a million dreams come to mind. It is exhilarating. However, the contrasts between the dream and the reality are in sharp focus on the streets of Hollywood themselves; the tacky and downtrodden shops; the tens of homeless, poor and sick; the casual violence in gesture and speech from the regular people walking along, unrecognised in its endemicity.

Eventually I find the other hostel I've been recommended and check in for one night as I'm low on cash. I would have changed my mind about staying if it wasn't my only option at this time of the day — the place is a dingy, dimly lit tip. The smell of paint stripper fills the halls. The walls and floors of the room I'm allocated are stained and damp, and three mattresses lying on the floor, as shoddy and ominously discoloured as the walls, are what pass as beds. The psychological impact of the locker machine for bag storage was probably the most disturbing aspect of the place. I reflected on all of the beautiful places and people I'd discovered in some of the poorest countries in the world. Now having travelled all around the globe and the worst place I've ever had to stay was found in the supposed pinnacle of human civilisation. The heart of the new world order. It was an irony that did not surprise me. I was not in a good mood, and the surly attitude of the incompetent staff did not help to improve it.

I got something to eat then went back to the hostel to attempt to sleep. Crossing the International Date Line had made my body clock whirr out of sync with L.A. time. I was feeling like a complete zombie but my nerves were too highly strung to risk unconsciousness. Another traveller enters the room and we warily reconnoitre one other in our first exchange of glances; a mutual attempt to ascertain the presence of a serial killer. He's a young Irishman abroad and we end up sharing a laugh over the state of our circumstances. Relieved, I can relax and sleep.

17/December

I get 'breakfast' in the kitchen. The place is truly a shithole of indescribable proportions; a derelict not fit for habitation; but at least the clientele are the usual backpacker kids, and I don't feel as threatened as I did when I first arrived. Everyone looks as horrified to be here as I am. The backpacking world is an incestuous thing, and I run into a girl from the *Wanna Taki*. A pretty girl with piercings and tattoos, I hadn't talked with her much on the boat as she'd been too busy getting it on with the diving instructor. Relieved to encounter at least a recognisable face we chat over tea and toast and then go on a walk in an attempt to find a bank.

Strangely, the banks are closed on a Sunday here and we are both in some need of cash. We manage to locate a credit service/money changer that cashes traveller's cheques. We're held up for ages by a guy at the front of the queue. A short guy sporting leather trousers and an open shirt and medallion combo, he whines and wheedles, doing everything but prostrating himself on the floor, as he tries to secure a loan. We eventually get served and access the precious funds that will keep us going during our time in L.A. Outside of the money changer I spot the short guy again. His character has inverted one hundred and eighty degrees; chest puffed out, bantering contemptuously with people passing by. He seems to think he is *The Man,* and wants to make sure everyone around knows. He jumps into a shining red Ferrari convertible and drives off at speed, burning rubber as he skids across the lanes.

A black hole of self worth orbited by glitz.

Me and the girl spend some time exploring around Mann's Chinese

Theatre and the walk of stars. Conversation had been slow and we make mutual excuses to go our separate ways. I'd resolved not to stay any longer in that damned hostel and eventually locate a place off the main boulevards that seems clean and reasonably priced. This dorm actually looks human, with four poster beds and newly washed and pressed sheets. I'm relieved to check in, and relieved to be on my own. My travel batteries were flat. All I felt like doing was resting up and taking stock.

I found a good little pizza joint on the strip for dinner, to treat myself to some greasy food and watch the world go by. This immersion in America had brought so much to mind. I had walked into a shop earlier in the day to buy a drink. I had the options of (i) Coca Cola, (ii) other soft drinks made by Coca Cola, or (iii) water, produced by Coca Cola. I had encountered the same phenomenon in London, St Petersburg, Moscow, Tokyo, Hong Kong, Beijing, Bangkok, Singapore, Sydney, Auckland and now here in Los Angeles. I grinned. I had to congratulate them. That was the sort of racket you want to be in on. Monopolies spanning an entire world. But don't believe the evidence of my globe-trotting senses; the new international economic order is all about: "Choice."

I looked out at the life on the streets of L.A. How similar to the life on the shining new streets of Beijing? The shops and the products and the clothes increasingly identical. The people subjects of a central government, remote physically and politically. Their lives beholden to a huge and paranoid external and internal security apparatus. But one was a totalitarian state, supposedly the very definition of modern evil, and the other "The Land of the Free and the Home of the Brave"?

I took respite from my musings in a show of *Borat* at Mann's Theatre. A biting satire; it had double the impact when sitting amongst a crowd of Americans.

19/November

A roommate arrives; Roeland from Cologne. An older man on holiday, he has the same expressive extroversion as Bastian. Must be something about that city? He seems a really good guy.

Today, the completion of a childhood dream: A visit to the La Brea tar pits, famed in palaeontology for their collection of Pleistocene flora and fauna (at this point the reader is not surprised that I was that sort of child). The exhibits were great. Well worth a twenty year wait, and that's not a joke.

In the evening watch *Back to the Future* in the hostel with Roeland. A wonderful film from my youth. For me, it summed up the best of American culture; the energy, the dynamism, the wholehearted exuberance, the outgoing sense of fun, the total optimism:

"If you put your mind to it, you can accomplish anything!"

It is easy to demonise the United States, for its post-WWII belligerence and its position as the driving force of corporate globalisation. America is a land of great natural advantages. But in their comfort and insularity the people had been lulled to sleep, and they had allowed their executive to be hijacked by special interest groups for the purposes of global domination and exploitation.

But I have a confession to make:

I love America.

Whenever I am here my overriding impression is of amazing goodness. Whether the natural wealth of the land or the spontaneous goodness of the vast overwhelming majority of the people. Reflecting on the positive contributions America has made to world society takes the breath away. Where would we be without their example of Freedom, Liberty and Democracy? Human Rights? In recent times, where would we be without their example of sexual equality? Or Martin Luther King's civil rights movement?

One man had a dream.

The dream came true.

That dream hadn't stopped on these shores; it had spread around the whole world and all of humanity is immeasurably better for it. Positive change in America, more than anywhere else, can ripple around the globe. I was conscious of the good in American society today. Whatever my criticisms of U.S. geopolitics or my initial impressions from the air, I had been enjoying myself immensely since I'd got here, hostel-from-hell excepted. Life on the ground in Los Angeles was great. So much to see and do. So much culture. So many amenities. Moreover, California was a tremendously creative place; my mind had been absolutely buzzing and my imagination on fire ever since my arrival.

I was inspired. To witness the power and the creativity and the achievements of American society is to be awestruck.

20/December

I head out of Hollywood to explore the centre of the city. There is a sizeable and picturesque old Chinatown and I enjoying browsing the shops and grazing the stalls. The most surprising discovery of all though is the remains of the original Hispanic town that gave the metropolis its name. The town square and an old Mexican-style Catholic Church still remain. I never expected such a charming heart to this sprawling city.

I felt at home in downtown Los Angeles. People were busy going about their commuting and their jobs just as they did in my own town. So many people, and such a wide cross section of humanity. Here in the story-telling capital of the world there were so many lives that went unreflected. Where were these people's stories?

They surely weren't anything like the escapist fantasies of "*The O.C.*", and they surely weren't represented by the exploitation of '*Gangsta Rap*'.

I got talking, or rather talked to, by a pair of the many borderline homeless that filled the city. A pair of panhandlers or mental health outpatients I didn't know which, they share the train journey back to Hollywood with me. One, the pusher of the wheelchair, looking much like a latter day Robin Williams, did the talking, whilst the other sat quietly in the wheelchair. After discovering I was a foreigner, he was quick to give me a State of the Union address. Exclaiming:

"We just don't make anything anymore!"

Referring to manufacturing and their economy.

"How can we when wages in China are $2 to our $10?"

He was laughing when he commented:

"How much does our government cost? How much do we give to the government and the IRS every year?"

Even more amused, he moved onto the many absurdities of modern life:

"How much are we spending on the war? Do you know how much we are spending on the war?"

"All the while we're racking up our national debt? Do you know what our national debt is now?"

I was surprised with the sharpness of mind behind the humble exterior. Moreover I was touched by the humanity of two people down on their luck supporting one another, and the good humour and ability to laugh about it all despite the circumstances. Their station arrived and Robin wished me a warm goodbye. The mute got up from his wheelchair and Robin sat down. The mute then wheeled Robin out in an exchange of roles from earlier.

Life is strange sometimes.

21/December

I took public transport from Hollywood to the airport. At one point some police entered the train and started checking everyone's tickets. I'd swapped lines a few times and I hoped I'd bought the right one. A Black man and his girlfriend were told to step outside the train. An officer looked at my ticket and also told me to step outside the train.

Oops.

Both myself and the couple were asked to sit on a bench surrounded by police officers. I was expecting the couple to be fined or asked to buy tickets but instead the man was forcibly pushed onto the ground by several officers, his hands cuffed behind his back, read his rights and searched. I was so shocked with the stereotypical heavy handedness that I almost started to laugh out loud. Only the other day Roeland and I had been making fun of the crazy U.S. policing shown live on TV, now here I was in the middle of it. It was unbelievable.

Another cop questions me but he lets me off once he discovers I'm a tourist on my way to the airport. I'm told to go buy another ticket at the machine downstairs. I walk along the platform and a Hispanic guy with a facial tattoo grins at me from among the crowd of onlookers and asks if I got fined. I tell him no and that they let me off with it. I'm about to commiserate with him over the situation but he starts fuming at the police:

"punk ass mutherfuckas, why I oughtta..."

I turn round and a cop is eyeballing me from the distance. I grasp the flammability of the situation. From a racial perspective it looks terrible: I'm the only Northern European in the whole vicinity (including the police) and I get let off with it, whereas the (probably homeless) Black guy with no $2 ticket gets the mass murderer treatment.

I buy a ticket from one of the machines downstairs. Some local kids see the money in my wallet and cluster round, trying to press me into buying sweets from them. I see the Hispanic guy heading down the stairs and he asks me to buy him a ticket too, under some impression that I owe him a favour.

"No"

"Why you punk ass mutherfucka, I'm gonna..."

He tails off into death threats as I headed back up the stairs. If death threats were at the top of this persons mind, I thought, what was at the bottom?

I waited at the end of the station for the next train, heart hammering and full of adrenalin. It had been a good lesson — however stupid situations may appear from the outside, they are completely real for those participating in them. The profusion of murder weapons in this society had made simple everyday situations positively incendiary.

I got on the train and had a look at the map in an attempt to try and work out where in the hell I was:

Hmmm. Sort of in the south of L.A. Sort of in the centre. *Ahh.*

I looked out at the passing scenery; an extreme case of urban desolation and decay. Empty lots, burned out cars and crowds of unkempt children playing on wasteland. In the distance were the glass towers of the business district.

The pyramid of human existence in stark relief.

Home

The plane takes off from the westward facing runway of LAX and the cloudless blue sky frames an extraordinary pacific sunset. As it banks around to fly eastwards Los Angeles comes into view. A vast cloud of brownish smog, which was completely invisible from inside the city, hangs wraithlike over the San Bernardino valley. Spilling between the crevices of the surrounding mountains in a concentrated brownness and staining the atmosphere far above a hazy yellow, the scale of this everyday pollution is monumental.

Another smooth flight. Thankfully there was some entertainment this time to break up the monotony. I was feeling more optimistic since my time in L.A. and looking forward to going home. People were beginning to wake up. In America, Australia and elsewhere. Change was in the air. People were beginning to question authority. What changes would I make in my own life?

I put the question to the back of my mind. It had been a long journey. Now I was just ready to go home.

Several sleepless hours later and Britain at last came into view. Owing to a quirk of December climate the sky was bright blue and cloudless at height. All of the clouds were situated on the ground in the form of a thick blanket of fog. It was quite a beautiful sight; the cloud so dense and so low that the contours of the English countryside appeared wrapped in flowing white satin.

The plane began its descent and the little stars greeted me before the bank of cloud. Once in the fog only the wings were visible; shuddering in the stress of the landing manoeuvres. Lower and lower the plane circled, still no sight of anything apart from the greyness was possible. Eventually we touched down — even on the ground the runway was barely visible, so thick was the fog! Spontaneous applause erupted from the passengers. The landing had been an absolute marvel of human technical ingenuity.

After a few hours I caught my second flight up to Glasgow. The fog gave way to broken cloud over Scotland and the landing was less dangerous this time. It felt so strange to be home. Familiar Glasgow airport now appeared so small compared to all the world scale hubs I had visited. The Scottish accents over the tannoy sounded like a cute provincial dialect rather than the way people actually spoke.

I hadn't informed my family of the exact date of my homecoming — just let them know before Christmas — in an effort to surprise them. Unfortunately my bank card had decided to stop working so I couldn't surprise them as best as I had intended. My brothers were still shocked when I phoned asking for a lift! I wondered if they had changed when I had been away, but when I saw them and hugged them they were exactly the same. It was only after a few minutes that I noticed a few wrinkles or a difference in weight to suggest the time away.

When I got home I did manage to surprise my mother and father, who were still mum and dad as ever. The strangest thing about the 20 month gap was the sameness. The standout difference I noticed in my household was the height of the trees in the garden. Just as the garden had grown outside the house, so had the amount of electrical cables and the amount of electronic toys within; the parallel growth of a cybernetic organism. My youngest brother had shot up like the trees in the garden. It was great to be back. It was almost Christmas.

PART II:
The Journey Internal

Mysticism De-Mystified

Mar/06 Malaysia

A *realisation came to me when I was sitting on a park bench in Kuala Lumpur. I was watching a man playing with his young child in a swing park, a scene of touching emotion and happiness that one can find everyday anywhere in the world. As I watched I was struck by the complete universality of existence. The life, the emotions were the same everywhere.*

I realised that the only place the universality of existence didn't exist was within the individual minds of people; even the very concept of Unity was itself interpreted through relative religious and cultural frameworks.

Gnosis

> *"A fragment of me in the living world is the timeless essence of life"*
> **– The Bhagavad Gita**

The unifying concept of all mystical traditions is *gnosis*; a term meaning knowledge of the divine light within. The path to *gnosis* is self-knowledge and meditation;

> *"The Kingdom of Heaven is within you"*
> **– Jesus**

> *"through oneself one may contemplate Oneself"*
> **– Lao Tzu**

> *"I am the inner sacrifice*
> *Here in your body, O Best of Mortals!"*
> **– Krishna**

"The Source is within"
– Rumi

"The Buddha is your own mind"
– Bodhidharma

Through development of his or her consciousness, the mystic eventually attains in deep meditation a non-phenomenal, non-conceptual awareness of existence as an undifferentiated ultimate reality[1]. This can be evidenced in the mysticism of all traditions:

Sufism

"there resides but one Reality above and beyond all forms"
– Sayyid Hossain Nasr

Yoga

"Entering a state of true illumination, nirbikalpa samadhi, these saints have realized the Supreme Reality behind all names and forms"
– Paramahansa Yogananda

Mahayana Buddhism

"The pure nature of Supreme Reality is the real Buddha"
– Hui Neng

These most profound experiences of enlightenment are ones in which there is no duality; no difference between subject and object, between the knower and the known[2].

"Rather, the (Father's) Kingdom is within you and it is outside you"
– Jesus

"As I am, so are others; as others are, so am I"
– Siddhartha Gautama

"I am the Truth!"
– Ibn Al Hallaj

"The Cosmos and I live together, Everything and I are one"
– Chuang Tzu

"When he perceives the unity
existing in separate creatures
and how they expand from unity,
he attains the infinite spirit"
– The Bhagavad Gita

Personal Spiritual Journey

My first momentous spiritual experience came when I was around 20 years old during a conversation with my girlfriend of the time when I was trying to describe to her why I believed in God. I loved my girlfriend very much and I searched for the words to explain to her what it meant to me; finally I could only say:

"God is Love"

As I said this what felt like an atom bomb went off in my heart. It felt like my heart had opened and that Pure Love or Pure God streamed out in an expanding explosion! It was singularly the most terrifying and powerful experience of my life and I clamped down tightly upon the sensation, shutting it out. Despite its momentous nature, it went completely unobserved to the world outside, even to my girlfriend it must have looked like but a momentary hesitation or pause for breath. I had never felt anything like it and I did not want to feel anything like it again so I forgot about the experience as quickly as I could...

...some years later I had reached a very low point in my life. My relationship with my first girlfriend had ended, my close circle of friends was breaking down as people moved out of town and on with their own lives and relationships, and as these important things disintegrated I was left with a life that consistent mainly of a job which was repetitive and uninspiring and which, simply put, I was growing to hate. At this time I was thinking a lot about the purpose of the universe, as is my contemplative nature, and I had decided to search for truth by first assuming the non-existence of God and then seeking to prove otherwise through scientific study. I searched for truth in physics, cosmology and in the background of my degree training; molecular biology and the related disciplines of genetics and evolutionary biology. However hard I tried and however much I read I simply could not find any coherent picture of Truth or existence of God.

One night as I was resting in bed, I realised that I had completely lost my happiness, my *joie de vivre*, my reason for living. As I thought about what had led me to this dire confluence of circumstances, I realised that it was simply:

me. My own decision making, or lack of. I had to face up to the reality I had created. I thought about what happiness meant to me, and an early memory surfaced of when I was a child of about four years old, thinking thoughts as I brushed my teeth and got ready for bed. It was my earliest memory of contemplating death. What would happen if mum or dad died? An emotional realisation that the whole world around me would collapse. It was their love that supported the life that I had. I realised that the atmosphere of my childhood had created the happiness in my life, and that in future I wanted to create a loving home for my own children. Love was the force that supported the universe.

As I was processing this reverie, the memory of that previous atom bomb experience returned from the deep subconscious depths of my being, as if to validate my new outlook. "The Kingdom of Heaven is within you". The answer had been there all along! I had just been looking in the wrong place. I realised I had previously been relying too much on my intellect and on second hand sources — from now on I would base my philosophy on the concrete of my own life experience. It followed that I would have to gain more life experience...

Not long after I began reading up on spirituality and meditation and one day I found the answer to explain that atomic occurrence: it had been an opening of the heart *chakra*. I was overjoyed with this discovery as before this I simply did not have the conceptual framework necessary to make sense of the happening. I found some *chakra* meditations and began to explore this new realm of understanding. I followed the methodology and remarkably found inner sensations of what I could only describe as *energy*. The more I discovered the more I became embarrassed about the depth of my own former hubris and arrogance; what did I think that very intelligent people on the other side of the earth were doing with their spare time for millennia?

I began to meditate daily and started to reap benefits immediately. I began to wake up in the morning not hating the world. I began to have more energy and more purpose in my daily life. As the months progressed I realised I was going through a process of *cleansing*. Memories, thoughts and emotions would drift into my consciousness at odd times as if for re-analysis. As well as the Hindu literature I had been reading, I also became attracted to *Zen* philosophy, with its emphasis on the everyday mind and enlightenment in an instant.

...one year later I had a chance encounter with a book, *Chi Kung (Qigong) for Health and Vitality*, written by Sifu Wong Kiew Kit, a *Chan (Zen)* master, while travelling in China. It inspired me to such a degree that I knew I had to try and learn from the master himself. I learned from the book as I travelled, and as I practiced the methodology I realised that I had found a complete system, much more powerful than anything I had come across before. At the beginning of the following year (2006), I had the opportunity to learn *qigong* and meditation from Sifu Wong himself. It was during this course that I had the experience which changed my world so completely...

Philosophy

Love of wisdom. Since the beginning of time, human beings have been looking out into the universe and asking — Why? How? What does it mean? Where are we going? Why are we here?

This exploration and questioning of the world around us gave rise to myriad worldviews and schools of thought. In the western tradition it coalesced into what was termed by the classical Greeks as philosophy. However there is one important difference between what constituted knowledge then, and what constitutes knowledge now; those ancient Greeks, like all prior peoples, recognised two directions of conscious investigation into the nature of things. That is, they not only accepted the external discovery of the world around that many of us nowadays consider as the only possible way to appropriate knowledge and understanding, but they also believed that knowledge could be gained from directing consciousness internally.

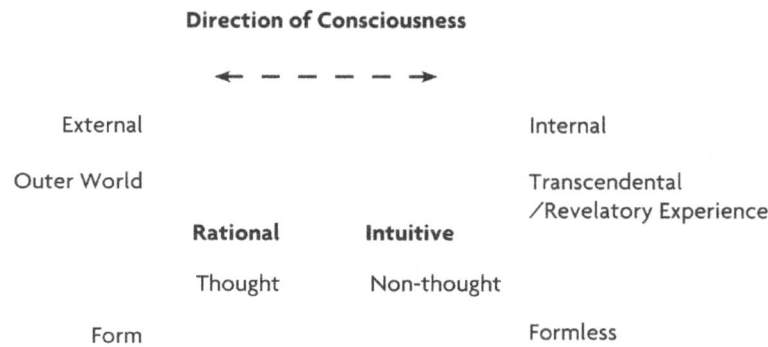

The above diagram eases conceptualisation of the mystical understanding of the universe, if one accepts the proposition that what has been termed transcendental or revelatory experience is possible.

As introduced in the previous essay, the philosophy expounded in this book concurs with the ancient worldview that internal direction of consciousness — and mystical experience developed from this — is possible. So before we continue further into a discussion of the nature of things, some definitions crucial to comprehension of the mystical worldview are in order:

Rational mind: *This is the aspect of mind reading, processing and assimilating the data from this page. This aspect of mind works according to linear logic. All inferential sources of data are processed through the rational mind.*

Intuitive mind: *Rather than concentrating precisely on the words on the page, if you instead notice how you are currently feeling — this brings you closer to your intuitive mind.*

Life experience: *If you take a moment to stop reading altogether, sit back, relax and just enjoy the world around you, and the flow of breath, in and out....*

The sensation of being comfortably alive, here and now, nestled at the centre of the universe; this totality of mind and body, as you know, is life experience.

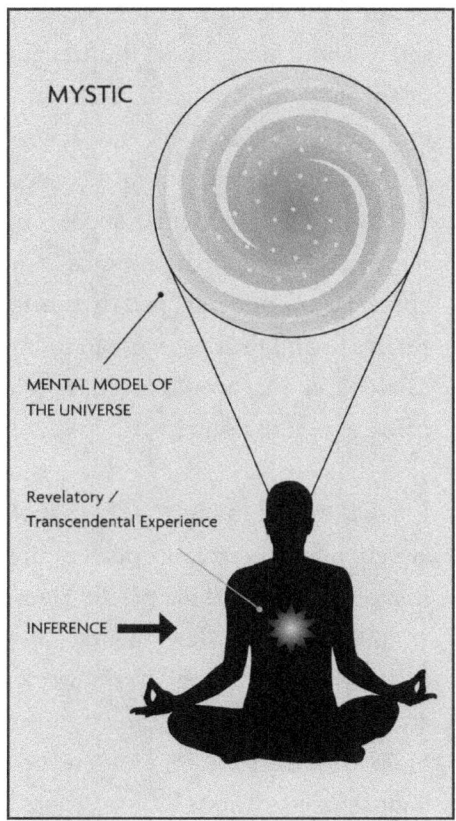

Mystical experiences do not belong to the rational world of book knowledge, although rational knowledge is involved in creating the mental frameworks necessary to understand and explore phenomenal real world information, and is also required to communicate this knowledge.

Mystical experience does not even belong simply to the realm of the intuitive mind; meditation practice enhances intuitive function and the linked rational faculties, but the flashes of conscious insight one may develop as one begins to train the mind are only the beginning.

Genuine mystical experience is much more profound — a quality of life experience itself.

The worldviews of the East, introduced in subsequent essays, stem from a mystical interpretation of the universe. The essential worldviews of the west began to diverge from those of the east with the influence of the Pythagoreans who introduced a mathematical and therefore rational emphasis to western mysticism.

In the cultural stream descending from those early Greeks, philosophy was divided into three main areas:

- Metaphysical
- Mental
- Physical

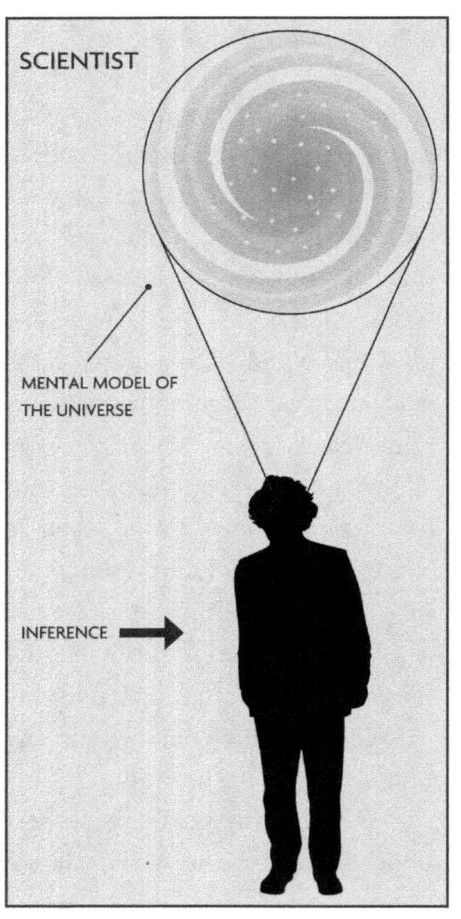

SCIENTIST

MENTAL MODEL OF
THE UNIVERSE

INFERENCE →

Over the course of the development of western society physics gradually became the domain of the physical sciences and metaphysics the domain of theology and religion, with contemporary philosophers endeavouring to straddle the ground between the two. The spectacular success of our endeavours in the physical sciences has of course shaped the world we live in and the lives we lead today to a degree that would be unimaginable to our ancestors.

Indeed, in our modern age, science largely plays the role in our culture that philosophy did in ancient times; as our modern day mental and metaphysical cultural frameworks are highly influenced by scientific biological, subatomic and cosmological investigations.

Our modern paradigm — that is, the beliefs, assumptions, perceptions and values that underscore our minds — is therefore primarily scientific and rational in nature. But underneath this modern worldview the ancient roots of western philosophy remain; an examination of these beginnings serves to shed much light on contemporary globalised world culture, and our own minds.

Roots of the Western Paradigm

"Let none who have not studied geometry enter here"
– Plato

The philosophic paradigm of classical Greece set the foundations for later Roman and Judaeo-Christian thought and as such represents the fundamental underlying worldview of western culture. The Greeks sought the first principles which gave order to a changing world, and thus separated absolute knowledge from the world around them, a separation which created a "Two World" understanding of the universe.

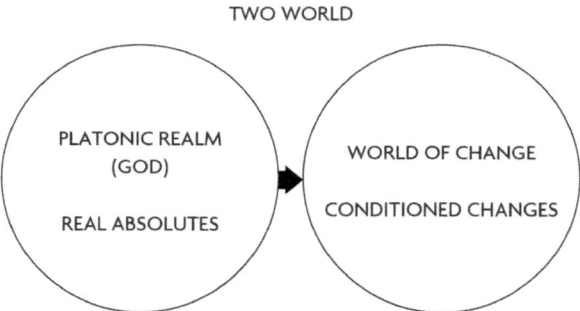

This primary distinction between Creator and 'the created', Order and 'the ordered' gave rise to the fundamental **dualist** categorisations of existence which we in the west can all identify with:

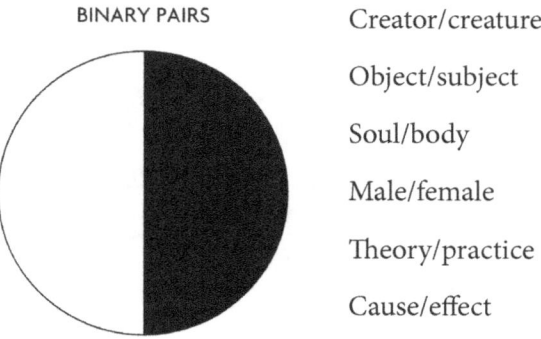

Creator/creature

Object/subject

Soul/body

Male/female

Theory/practice

Cause/effect

A crucial quality of these binary pairs is that those belonging to the realm of absolutes are considered to be *independent* and *superior* to those of the world of change, which are inferior and dependent on the first. These root assumptions give rise to many features of our paradigmatic background:

Causality

The assumption of separate and pre-existent order predicates an original beginning (cosmogony) and a final purpose (teleology) of the universe according to a linear line of causality.

Objectivity

The assumption of Two Worlds provides the conceptual basis for the traditional western stance of objectivity — the position of separating ourselves from the world and viewing the totality of existence from an external perspective. From this viewpoint, positions in the field of existence can be de-contextualised as "objects".

Purpose

The purpose of human existence becomes one of discovering and conforming to a universal pre-existent design which forms the basis of order; both natural and moral.

Knowledge

Mathematics, belonging to the realm of absolutes, became the basis for knowledge, as illustrated by the above quoted plaque that hung over the door of Plato's academy. As such, categorisation of the world became analytical and abstract; a quest of breaking down and identifying the universal structures behind changing phenomenal appearances.

Creativity

Creation and creativity, as qualities of the pre-existing platonic realm, are identified with design and origination.

Socio-political Order
Socio-political order is understood as compliance to a universal standard independent of oneself, as with a Constitution, a Bill of Rights, or a Commandment of God.

Flaws of the western paradigm

Aside from the obvious asymmetry of dualist categorisation there are two stand-out points in reference to modern scientific thought:

- **Objectivity**

Our common discursive stance places abstract theory in the realm of absolutes rather than as relative inter-subjective assumptions. This produces the risk of mistaking our conceptual maps for the territory of reality.

- **Reductionism**

A combination of an external, objective stance and its associated de-contextualisation along with linear causality produces the reductionist character of historic scientific analysis. An understanding of wholes is extrapolated from an understanding of parts.

An Introduction to Buddhist Philosophy

There are two extremes, monks, which are to be avoided. What are these two extremes? A life given to pleasures, dedicated to pleasures and lusts — this is degrading, sensual, vulgar, unworthy and useless; and a life given to self-torture — this is painful, unworthy and useless. By avoiding these two extremes, the Perfected One has gained the knowledge of the Middle Path which leads to insight and wisdom, which produces calm, knowledge, enlightenment and nirvana.
– Samyutta Nikaya V 420[1]

Historical Buddhism originated from the teachings of Siddhartha Gautama (400 BCE), known as the Buddha. In his first public sermon, the *Turning of the Wheel of Dharma*, he expounded the core of Buddhist doctrine known as the Four Noble Truths:

1. *Birth, old age and death are unsatisfactory (duhkha).*
2. *The cause of unsatisfactoriness is craving (trsna).*
3. *There is an end to unsatisfactoriness and craving and it is known as nirvana.*
4. *The way to achieve nirvana is to follow the middle path between all extremes. This Path has eight steps and involves the development of 1. Appropriate view 2. Appropriate intention 3. Appropriate speech 4. Appropriate action 5. Appropriate livelihood 6. Appropriate effort 7. Appropriate mindfulness 8. Appropriate meditative concentration.[2]*

Acceptance of The Four Noble Truths, together with the moral law of karma, forms the basis of Buddhist practice and teaching. To comprehend this practice and teaching it is necessary to understand the Buddhist view of reality and what the Buddha termed the three marks of existence (*tri-laksana*):

- Suffering (*duhka*)
- Impermanence (*anitya*)
- No-Self (*anatman*)

These three marks of existence follow one from another — while we crave for things we cannot achieve true happiness; any states of happiness that we do manage to achieve are bound to pass away through the impermanence of changing conditions or the course of time. This inability to maintain the universe as we desire is the cause of all frustration, and ultimately suffering. As all things are found to be impermanent then they must also lack a permanent basis, or self, of their own.

To put the Buddhist worldview into its proper context, one must be aware of the metaphysical postulates of the time; existence in the contemporary Brahmanic and Jaina traditions was understood as a series of births and deaths (*samsara*), with the aim of liberation (*moksa*) from this cycle. Although aspects of this metaphysical scheme remained in Buddhist teaching, that is, existence as a series of rebirths (*samsara*) until liberation (*nirvana*) is obtained, the doctrine of no-self was a radical departure from the Brahmanic religious teachings of that era, which were based on the concept of self (*atman*).

The Buddha provided an explanation for this radical departure with the doctrine of inter-dependent-origination (*pratityasamutpada*). As opposed to the postulate of an abiding substance (*atman*) as a source of order and existence, inter-dependent-origination describes existence as a consequence of the dynamics of the universe, conditioned by causal relationships.

Sunyata

The historical high point of Indian Buddhist philosophy was developed by the monk Nagarjuna (2 C.E.) of the Madhyamaka school of Mahayana Buddhism. Nagarjuna unified several Buddhist positions with his exposition of the doctrine of emptiness (*sunyata*); the concept that all things are empty of independent existence. This combination of no-self and inter-dependent-origination created an understanding of the mutual relativity of all things.

According to the worldview of Nagarjuna, you now hold sunlight in your hands*. Without energy from the sun there would be neither photosynthesis nor a water cycle necessary to nourish, in combination with soil nutrients and air, the tree from which the paper of the book originated. Without sunlight there would be no grand global metabolic flow necessary to produce all of the foodstuffs needed to nurture all of the people — everyone from the writer, to the loggers, hauliers, publishing staff, distributers, shop assistants, and so on — involved in its production. If we isolate any one of these numberless constituents we can see that any particular factor we identify is in turn equally dependent on other conditions for its own existence; the human beings all dependent on the existence of their parents, those parents in turn were dependent on the existence of their own parents, and so on, *ad infinitum*. The book you are holding can therefore be understood to lack any own-being; to be empty of any independent existence.

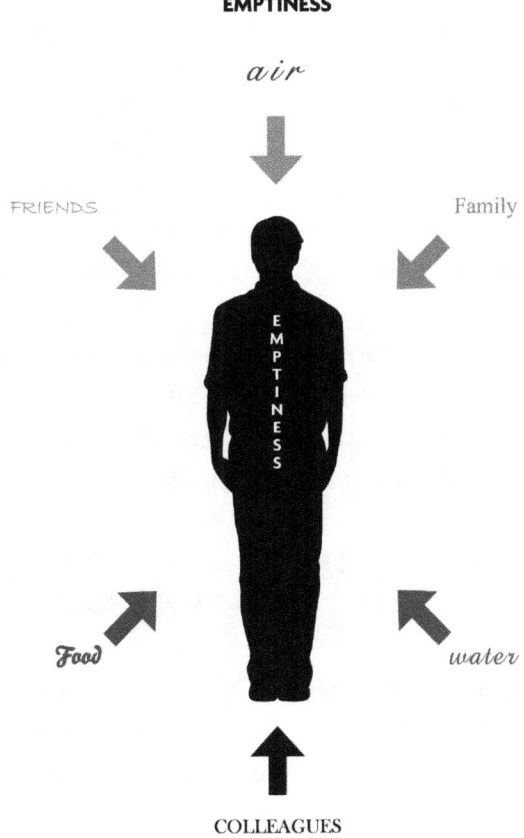

Even the very concept of the book is relative and dependent on conditions; another person unfamiliar with the language in which it is written would not interpret it as you do, a pet dog watching you read would see shape and contrast in your hand but no "book", to a bacterium dwelling on top of it the book would represent a very different sensory experience indeed! All things are conditioned according and relative to the mind of the observer in other words. The mind of the observer itself is completely dependent on conditions (such as the book in front of it) and is ever changing, therefore no single static entity can be found to exist anywhere in the entire universe! All things observed with the perspective of *Sunyata* will unravel into a set of interdependent causal relationships.

There is no self in Buddhism because the only real "Self" that can be found to exist from the Buddhist perspective is the entirety of the universe; past, present and future. There is nothing that can be found that exists in isolation, and there is no first cause that can be established, only an infinite web of conditional relationships. "Emptiness" therefore is not nothingness. Instead it is an understanding which moves us from the conventional, habitual and everyday worldview of individual people, objects and events into a comprehension of existence as composed of dynamic processes and webs of relationships; an understanding in which the relativity of subject and object is always underscored.

To opponents who interpreted his deconstruction of their own postulates as nihilism, Nagarjuna reminded them of the Buddhist doctrine of "two levels of truth". Individuals, objects and events are conventionally real but *sunyata* serves as a pointer to ultimate truth which must be **directly experienced**.

> *"Form is Emptiness, Emptiness is Form"*
> **– The Heart Sutra**

As you will no doubt find from any time spent contemplating on it, the importance of the Buddhist worldview cannot be understated. One of the central metaphors is that the nature of ultimate unity is phenomenal difference; existence is inherently dynamic and composed of ever increasing variation and multiplicity.

*The example holds for an electronic version; replacing logged paper with mined electronic components!

Jnana

Buddhism is often misinterpreted due to its subtlety and depth; indeed there are multiple positions within the sects of Buddhism itself. The core doctrines paint a very negative picture of existence when interpreted with the rational mind alone. The fact that many historic and contemporary western academics and intellectuals have interpreted Buddhism as radical scepticism and nihilism is more illustrative of the perceptions of those particular western academics and intellectuals than it is revealing about Buddhism, for example.

In order to comprehend this teaching, it is helpful to consider the Buddha's explanation of his own position; he gave his followers council as a spiritual physician and described his teachings as a raft to reach the other shore (*nirvana*). The Buddha's prescription for overcoming suffering was a life of ethical behaviour and the seeking of direct *gnosis* through the training of the mind: *appropriate effort, appropriate mindfulness, appropriate meditative concentration.*

As we have learned in the previous discussion, enlightenment experiences are those of non-duality. The three marks of existence can be reinterpreted from the perspective of these most profound of experiences:

- Suffering – as relative to the bliss of *nirvana*
- Impermanence – as a root descriptor of the nature of reality
- No-Self – as knowledge of non-dual experience preserved in the world as a philosophic and doctrinal position

Buddhism is not an exercise in persuading yourself that you don't exist, rather an understanding that you are a process that exists only in relation to other things. Its ultimate aim is not the extinction of existence, but the extinction of suffering. It has been described by many commentators as a religion without God. However, I would say that God is centrally present in Buddhism, more so than in any other religion that I am aware of. The outside God in Buddhism is an understanding that you are not separate from a universe defined by interdependence. To realise the inside God is the aim of Buddhism; direct *gnosis*. God is present in Buddhism, but hidden under markedly less corruptible terms such as *nirvana*.

For example; *nirvana* resists conceptualisation by the rational mind and is found only through self-development and meditation. *Nirvana* is a direct experience attainable in the here and now. *Nirvana* does not have the qualities of personhood but it is realised by individuals. If you wish to spread *nirvana* you must first spread compassion, peace and happiness. It cannot talk to the leaders of one's society and it is extremely difficult to use as a justification for an offensive war.

Buddhism could be described as the mystical path enshrined in society rather than as a religion in the sense of the near eastern traditions. The central scriptures always seek to remove the conventional seeking for an outside God and instead point directly to the mind of the individual:

*"Our nature is the mind. And the mind is our nature. This nature is the same as the mind of all buddhas. Buddhas of the past and future only transmit this mind. Beyond this mind there's no buddha anywhere. But deluded people don't realise that their own mind is the buddha. They keep searching outside. They never stop invoking buddhas or worshipping buddhas and wondering **Where is the Buddha?** Don't indulge in such illusions. Just know your mind."*

– Bodhidharma[3]

This also explains why Buddhism, aside from accepting a causally connected cycle of death and rebirth, lacks much of the metaphysical speculation found in other traditions — because the answers to all of the "Big Questions" have already been pointed to: they are found internally.

Classical Chinese Paradigm

"Solitary Yang prevents Life, Solitary Yin prevents Growth"
– Inner Classic of Medicine

In contrast to the origins of western thought, the classical Chinese paradigm begins with the root assumption that there is only one world which
is both the origin and field of all existence.

ONE WORLD

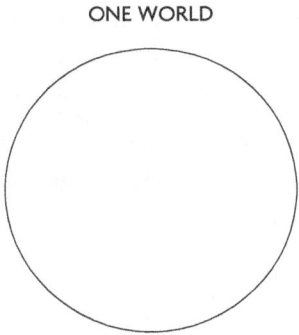

One World

Creativity and order are thus immanent in the world itself rather than externally activated from without. Order and change are equal realities and a quality of things-in-themselves.

The classical Chinese conceived of this One World as a Cosmos of *Qi*
(pronounced: Chi) — energy which is simultaneously psychic and physical
and which manifests in various patterns and dynamics. Understanding of this
one world is traced out from that which is most primary and immediate; the
perspective of any one individual. As all objects and events are perceived from
a certain perspective they are related, conditional and continuous to the position that perceives. All things are therefore a condition and a relation to every
other. From this viewpoint, 'events' and 'objects' are understood as temporal
processes and sets of relationships.

Category is relative and contrastive, as illustrated by the yin-yang polarities which ground Chinese thought:

Yin-Yang

Passive-Active
Female-Male
Night-Day
Cold-Hot
Winter-Summer
Lower-Higher

These polar opposites differ fundamentally from the western conception: they are symbolic, relative and dynamic rather than absolute and static. For instance; spring is yang to the yin of winter, but then becomes yin to the yang of summer. They are *connected* rather than separate, *interdependent* rather than independent, and together represent an *implicit unity* rather than an explicit opposition. All change in this one world is understood as movement on a continuum between such polarities.

Causality

Without a root cause/effect distinction, all things in the field of existence are mutually acting and being acted upon. The way of things is understood as *Tao* — a pattern mapped from a specific perspective; an experiential pathway traced through existence that defines both the how and the why.

Objectivity

As there is only one world it is not possible to separate oneself from it and then 'look in from outside'. All objects in the field of existence, including ourselves, are found to be unique, contextual and a function of a network of relationships. Classical Chinese thought therefore lacks the formal abstraction of the western paradigm and the complex theoretical structures which arise from this.

Purpose

The purpose of existence is immediate rather than abstract, intrinsic rather than determined; to coordinate the myriad details of the here and now in an effort to create optimal sets of conditions. The secret to this best use of circumstances is understood as harmony — the creative balancing of the various yin and yang components of life.

Knowledge
In contrast to the west, knowledge is synthetic and experiential rather than analytic and theoretical. Knowing and doing are interlinked; a mapping of patterns within the dynamic processes of life which, once ascertained, can be manipulated to improve conditions. Communication becomes the paradigm for knowledge, and the proviso of the wise. Indeed the very characters of Chinese language itself are dynamic, representing a complex of emotive associations rather than being precise symbolic abstractions as in the western tradition.

Creativity
Lacking the underlying abstract assumptions of the western paradigm, order and creativity arise from the interaction of the various components of the world over the course of their existence. Creativity for the classical Chinese begins from one's own place and is synonymous with the experiential journey of life; understanding and then coordinating conditions in an ongoing effort to cultivate the most productive and harmonious sets of circumstances.

Socio-political Order
Each individual is understood as the centre point of a variety of social roles, relationships and responsibilities. Order then begins with oneself and radiates outwards. The cultivation of social harmony is therefore implicit in the cultivation of personal harmony. Classical Chinese social order is thus interdependent and symbiotic; a "bottom-up" and emergent phenomenon.

Holism
The classical Chinese paradigm is holistic; parts are understood within the context of the whole in which they take place.

Wu-Xing
From the assumption of one world and an understanding of the interconnected nature of phenomena implicit in holism comes *Wu-Xing* or the Five Phases.

The classical Chinese thinkers studied the changing patterns of phenomena in the world and categorised them into five archetypical processes of transformation. These five processes were considered to endlessly inter-create and inter-destroy one another in the operation of sustaining the Cosmos. When the Cosmos was in a state of yin-yang harmony, the inter-creative and inter-destructive processes were considered to be in balance.

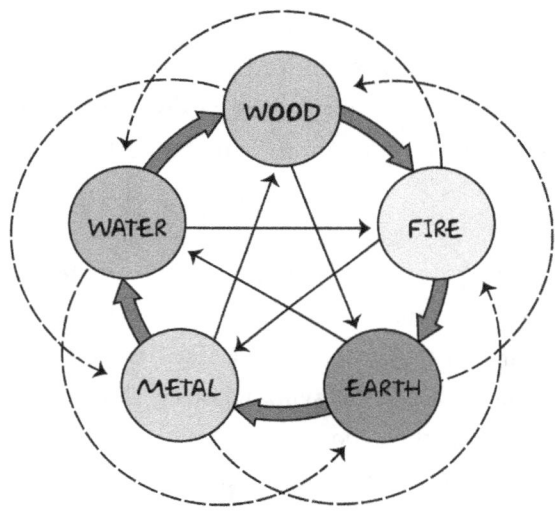

The classical mnemonic for inter-creativity is:
Wood creates Fire. As a wooden log is consumed in flame. Fire creates Earth. As ashes are left over from a bonfire. Earth creates metal. As metal is dug up from under the ground. Metal creates water. As metals liquefy when heated. Water creates Wood. As trees grow when watered.

The classical mnemonic for inter-destructivity is:
Wood destroys Earth. As trees take up the essence of the soil. Earth destroys Water. As soil can be used to fill in a pond. Water destroys Fire. As water can be used to put out a blaze. Fire destroys metal. As a furnace melts down iron. Metal destroys Wood. As a metal axe can be used to chop down a tree.

As there is no root cause/effect distinction, all of the processes of Five Phases are considered to be constantly acting and interacting with one another simultaneously. It must be noted that these Five Phases are symbolic rather absolute. As such, they can be used to describe both (i) the character and (ii) the patterns of interaction of processes observed in nature:

- Wood represents a growing, elongating process
- Fire represents a consumptive, transformational process
- Earth represents a centring, stabilising process
- Metal represent a resonant, catalytic process
- Water represents a fluid, expansive process

So far, so alien. Real world applications may help to bring a radically different paradigm into sharper focus.

Wu-Xing and the human body
Without the precise animate/inanimate differentiation of the western tradition, patterns of change in the macro cosmos of the outer world are equally applicable to the micro cosmos of the inner world of the human being.

Thus, the Five Phases in traditional Chinese medicine are applied to the major organ systems of the human body:

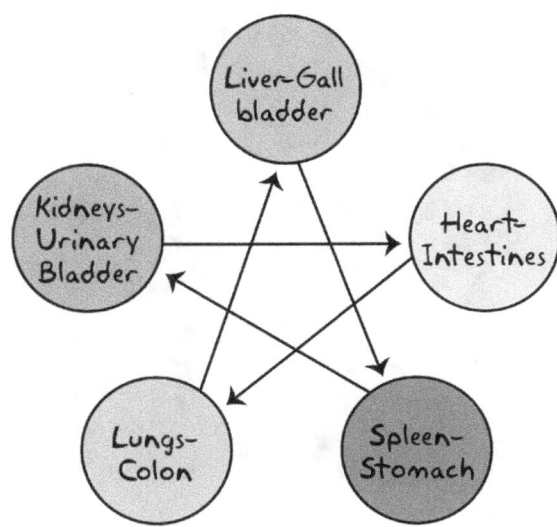

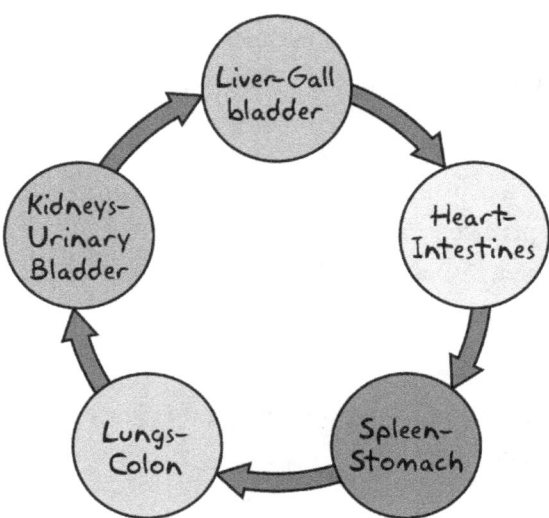

Again, lacking a theoretical soul/body distinction, they are also applied to the major emotions of the human psyche and their relationship to the major organ systems:

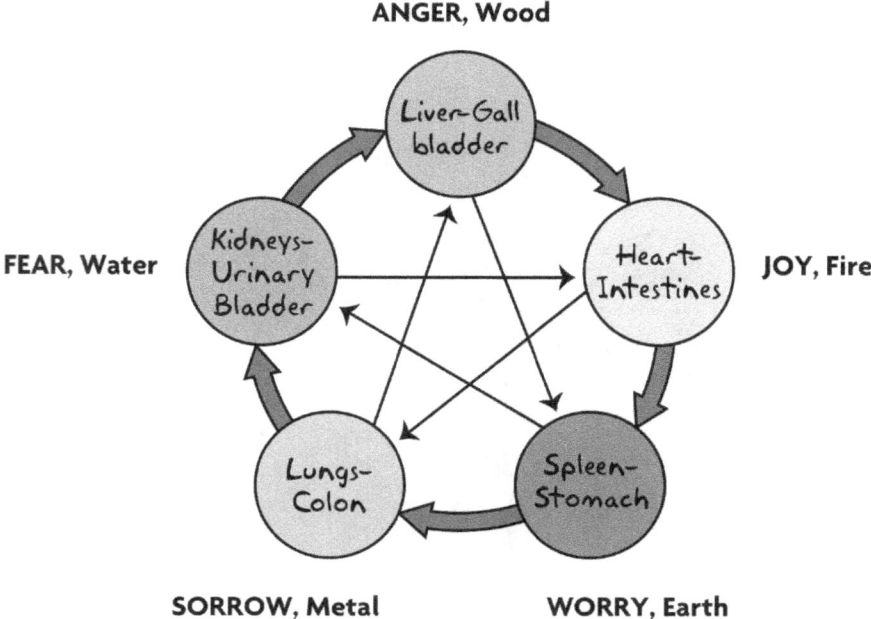

With this understanding, a traditional physician is equipped with a means toward mapping the state of the whole system of the human being through comprehension of the activating and controlling relationships enabling its natural self-regulation, and thus has a basis for (i) identifying disease, and (ii) the treatment of disease through fortification and stimulation of interacting parts of the human body.

For instance; a patient presents with symptoms of proneness to anger and reduce appetite. This is expressed as: "Wood (liver) destroying Earth (stomach)". The physician may therefore treat the patient by fortifying his liver rather than attending directly to his stomach.

In this manner, the traditional physician is able to ascertain disease locus rather than only symptom locus, with the corresponding ability to cure the root cause of the ailment.

Another example; a patient presents with symptoms of tuberculosis. The physician provides a treatment regime of nutrient rich food along with appropriate medication to increase appetite and aid digestion. The aim is an increased energy supply to the patient's body, which will flow to the lungs and enhance the natural ability of the body to repel invading organisms. This is expressed as: "Earth (stomach) creating Metal (lungs)".

In this manner, the traditional physician is able to treat a particular disease locus by fortifying a related organ system.

It should be noted that, in contrast to the genesis of many hypotheses in western thought, the idea of the Five Phases did not come before observation of the inter-sustaining nature of the human body systems; the inter-sustaining nature of the body systems was first observed, and then the concept of the Five Phases was applied as a descriptor of their patterns of interaction.

When applying the Five Phases to the human body, it must also be held in mind that Wu-Xing, like yin-yang is a symbolic and relative understanding, rather than an independent, originative, absolute and static theory structure; a useful guide on the pathway through complex phenomena rather than an always precise and reproducible descriptor of reality.

Creating a New Paradigm: An Ecological Paradigm

Science/Gnosis

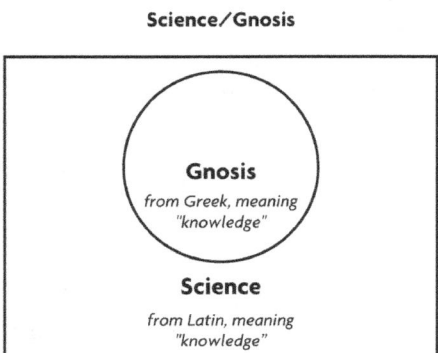

	Rational	Intuitive
Position	Objective	Perspective
Entities	Independent	Interdependent
Logic	Linear	Nonlinear
Knowledge	Analysis	Synthesis

From the contrast of the preceding paradigmatic overviews we can see the western tradition as an example of an externally focused paradigm built on the foundations of the rational mind, with the Australian Aboriginal paradigm representing an example of a primarily intuitive basis and the classical Chinese paradigm occupying a relative middle space between these two extremes.

Buddhist philosophy includes many complex and advanced rational concepts, yet it is also intuitively based. **Sunyata** is not an idea developed from intellectual speculation, but an understanding of reality based on direct revelatory/ transcendental experience.

On reflection on these world-shaping paradigmatic foundations, it becomes clear that the now globalised western tradition gives rise to a limited and unbalanced set of beliefs, assumptions, perceptions and values in relation to the world around us.

A new paradigm is required; to develop this we can learn from the paradigms of the east and those of the Aboriginal populations of the world. In contrast to the current overriding worldview, this will be a paradigm which understands the dimensional limitations of the rational mind, and which recognises the primary importance of non-thought and internal exploration — *gnosis* — in human existence and understanding.

"If you use your mind to study reality, you won't understand either your mind or reality"

– Bodhidharma

This new paradigm is a change to Science — no longer the premier route to truth; the essential limitations of the scientific method must be culturally recognised; theory is not transcendent truth but relative inter-subjective ideas, and experiment can never completely describe reality:

When scientists use experimentation attempts to describe a phenomenon, they must remove the entire rest of the universe from the calculations of their controlled experiment. As such, experimental results can never completely reflect the reality of the phenomenon in question.[1]

As an antidote to this, with **Sunyata** we can see that to describe any phenomenon fully, we must trace out its relationship to the entire universe.

Science should be agnostic; as gnosis comes from a different faculty of mind. Note that these two approaches only conflict if the rational mind blocks the intuitions — however with both modes of discovery, enhanced understandings can be realised.

"If you study reality without using your mind, you will understand both"
– Bodhidharma

This new paradigm is a return to pre-eminence of Gnosis; the route to answering all of the "Big Questions" for oneself. In this new worldview, science and gnosis will be the relative knowledge systems mapping the journey through the outer and inner experience of life, respectively. Both modes of consciousness ultimately parallel and complement the other.

These changes do not mean abandoning the strengths of the western paradigm, without which science and many other aspects of life could not have advanced to the degree they have done over the centuries, rather combining these strengths with the strengths of the intuitive paradigms and their greater understanding of participation in living systems.

A new paradigm is born transcending the weaknesses of either emphasis; an ecological paradigm whose central aim is the cultivation of harmony, based on an understanding of the interdependent nature of reality.

Where is this Unity?

If the mystical traditions of the world all agree on the Unity of All Things, then why are there so many different interpretations of this?

Rational Mind:
The purpose of the rational mind is differentiation and evaluation. It functions by separating the field of existence into objects, and it does this according to linear logic. Linear logic is two dimensional by its very nature.

As an example, in the realm of terminology consider the Hindu atman (self) with the Buddhist anatman (noself).

These are clearly logically mutually exclusive terms.

When functioning from the rational mind anything we can define is defined by its difference to something else. Moreover, the nature of phenomenal reality itself, as described by **Sunyata**, is inherently dynamic and composed of ever increasing differentiation and complexity. Every mind is a worldview and each perspective is unique.

Intuitive Mind:
The purpose of the intuitive mind is synthesis and coherence. It functions by grasping contexts and relationships. Its logic is nonlinear and multidimensional.

In its historical context, Buddhism can be seen as a conceptual revolution of the foundation; a movement from a dualistic "Two World" to a non-dual "One World" model of existence. It is difficult to locate precise concepts in a tradition that seeks to cut its members free of ontological speculation, but an equivalent to the metaphysical Hindu term *atman* exists in Buddhism in the form of *tathagatagarbha*; buddha-seed or buddha-essence.

"In the body of mortals is the indestructible Buddha nature. Like the Sun its light fills endless space. But once veiled by the dark clouds of the five shades, it's like a light in a jar, hidden from view."
– The Sutra of Ten Stages

All Self (*atman*) and No Self (*anatman*) philosophies can therefore be seen as two different ways of expressing the Unity of All Things.

Life Experience:
Moving our focus further out we gain an even better perspective on the universality of life.

For instance, consider a Japanese Zen Buddhist attempting "to see his Original Face" and an Indian Yogi attempting "to attain Union of Atman with Brahman". We can see that these are two human beings using the same or similar methods to reach the same experience. Meditation quiets the rational mind and allows consciousness of Unity to come into Awareness. This experience of Unity is then filtered back into the world through relative cultural and philosophic frameworks and according to the phenomenal abilities of the individual mind experiencing it.

Metaphysics II

"Experience is the product of the mind, the spirit, conscious thoughts and feelings, and unconscious thoughts and feelings. These together form the reality that you know."

– Seth

A spiritual underground movement has been growing in the west in recent decades which reintroduces a very ancient esoteric concept known as *The Law of Attraction*. According to this philosophy of Idealism, there are three fundamental *Universal Laws*:

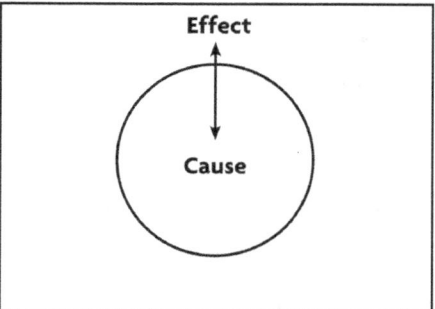

1. The Law of Attraction

2. The Law of Intentional Creation

3. The Law of Free Will

This is a similar existential scheme to that of the earlier described esoteric model of *karma*, however, the concept of *karma* can be misinterpreted as the fatalism of being eternally subject to past events, or of the
continual labouring under a moral sum.

In this alternative model the point of power is in the present, and
the emphasis is on what the individual sows and reaps moment-to-
moment, emotion-by-emotion and thought-by-thought.

Further Reading
Ask and It Is Given by Esther Hicks
The Nature of Personal Reality by Jane Roberts

The Death of God in the Mystical Traditions

The historical effect of spiritual revelation and the social changes which proceed from these messages is not a common point of discussion in our histories and philosophies.

Take for instance the changing cosmologies introduced by these alternative spiritual movements:

Brahmanic Orthodoxy:

In the beginning this world was just a single body (atman) shaped like a man. He looked around and saw nothing but himself. The first thing he said was, "Here I am!" and from that the name I came into being...He wanted to have a companion...So he split (pat) his body into two, giving rise to husband (pati) and wife (patni)...He copulated with her, and from their union human beings were born... It then occurred to him: "I alone am the creation, for I created all this."

*– **Brhadaranyaka Upanisad***

Buddhist Alternative:

After a world system has passed away, a new one evolves. The first being that appears then thinks:

"I am the Brahma, the Great Brahma, Conqueror Invincible, Seer of All, All-powerful, the Lord, the Maker, the Creator, the Noblest of All, Assigner to each his station, accomplished in attainments, the Father of all past and future beings."

*– **Brahmajala Sutta***

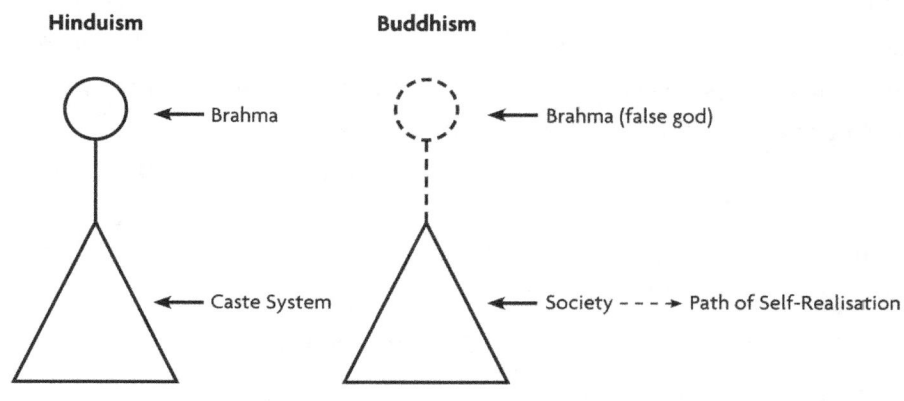

Hinduism   **Buddhism**

← Brahma   ← Brahma (false god)

← Caste System   ← Society - - - → Path of Self-Realisation

Christian Orthodoxy:

"I am the LORD thy God, which have brought thee out of the land of Egypt, out of the house of bondage. Thou shalt have no other gods before me."

"And God said, Let us make man in our image, after our likeness: and let them have dominion over the fish of the sea, and over the fowl of the air, and over the cattle, and over all the earth, and over every creeping thing that creepeth upon the earth."

– Exodus 20.2–3, Genesis 1.26

Gnostic Alternative:

Their chief is blind; because of his power and his ignorance and his arrogance he said, with his power, "It is I who am God; there is none apart from me." When he said this, he sinned against the entirety. And this speech got up to incorruptibility; then there was a voice that came forth from incorruptibility, saying, "You are mistaken, Samael" — which is, "god of the blind."

The rulers laid plans and said, "Come, let us create a man that will be soil from the earth." They modelled their creature as one wholly of the earth."

...and the rulers gathered together all the animals of the earth and all the birds of heaven and brought them in to Adam to see what Adam would call them, that he might give a name to each of the birds and all the beasts.

– The Hypostasis of the Archons

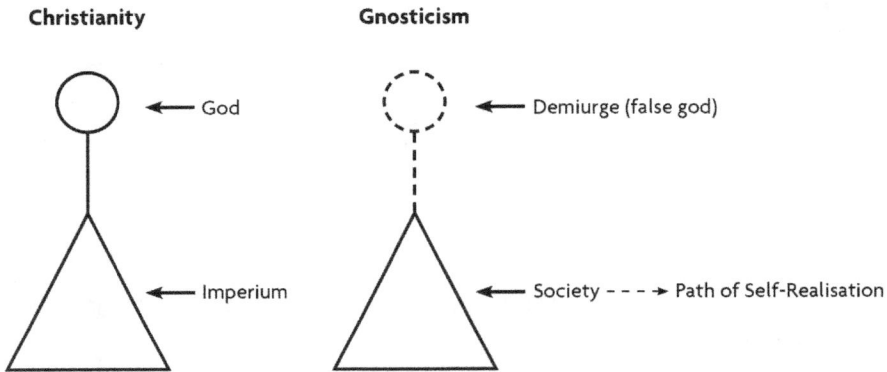

There are a number of important reasons behind these differing accounts of creation however, to cut straight to the philosophic heart of the matter — to the mystics of Buddhism and Gnosticism the competing hegemonic scriptures provided their own self-falsification as they promoted God as an external phenomenal entity, rather than an internal and directly experienced ultimate reality.

Mind Training

"Hard it is to train the mind, which goes where it likes and does what it wants. But a trained mind brings health and happiness. The wise can direct their thoughts, subtle and elusive, wherever they choose: a trained mind brings health and happiness."

– Siddhartha Gautama

Meditation need not be thought of as esotericism, and it is not about self-denial; it is about health and happiness.

"Right attention follows from right effort. It means keeping the mind where it should be. The wise train the mind to give complete attention to one thing at a time, <u>here</u> and <u>now</u>."

– Siddhartha Gautama

Meditation in daily life is mindfulness; *Zen* is keeping one's mind in the present moment. To focus the mind in the here and now is to cultivate the optimal state of being; one most familiar to musicians, athletes or artists engaged totally in their performance. Mindfulness allows full experience, full expression and the cultivation of excellence in all areas of life.

"Whatever is positive, what benefits others, what conduces to kindness or peace of mind, those states of mind lead to progress; give them full attention. Whatever is negative, whatever is self-centred, what feeds malicious thoughts or stirs up the mind, those states of mind draw one downwards; turn your attention away."

– Siddhartha Gautama

Focusing the mind at one point is the initial aim of mind training in many mystical systems. One-pointedness of mind (one thought) is a necessary precursor to achieving states of zero-mind (non-thought) in deep meditation.

What is the purpose of achieving a state of non-thought? An empty mind is the gateway of revelation.

"To the mind that is still, the whole universe surrenders"
– Lao Tzu

The importance and power of our own consciousness becomes clear as the practice of meditation training progresses. This underscores its proper use:

"Avoid all evil, cultivate the good, purify your mind: this sums up the teachings of the enlightened ones."
– Siddhartha Gautama

Spirituality in the west is not often associated with power; however to train the mind necessarily increases ones capacity for action and therefore responsibilities; thus the central importance of ethics in all traditional systems of meditation.

Ignorance

In the mystical traditions of the world, the problems of human existence are understood in terms of ignorance rather than sin.

"May all beings be happy"

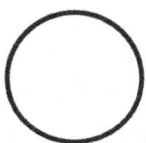

Jesus	Thich Naht Hanh
Mevlana Al-din Rumi	Bertrand Russell

"If you control the food you control the people, if you control energy you can control whole continents, if you control money you control the world"

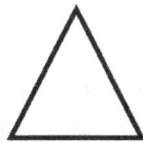

Cesar Borgia	Pol Pot
Osama bin Laden	Josef Stalin

Spirituality is often considered an issue of belief and belief system, however if we look at the examples of lives considered to be spiritual, we can recognise a common **State of Being**.

    If you can for a moment put aside the argument that Bertrand Russell would no doubt start if the first four people were put into an imaginary room together, we can see a similar state of being shared between these representatives of very different beliefs. These people all developed their own minds. They all cultivated and put into practice a sense of universal love. We could define their philosophy as:

Love of life.

The second group also share broad similarities. These people cultivated their own sense of self-aggrandisement. They cultivated the aggrandisement of what they considered as their own greater selfhood. To these ends they sought to dominate, exploit or destroy 'The Other'. All of these people were political leaders. We could define their philosophy as:

Need for control.
In these two groups we find a number of different individuals with diverse belief systems all relative to their particular place and time. While these are perhaps extreme examples, the contrast does serve to shed light on the potentials of human nature and free will.

The second group is composed of demonised figureheads, demonised in part because they ended up on the wrong side of the writers of history. Of course the grand abuses of human history — such as the abuses we associate with a name like Josef Stalin — were not interpersonal. They were social; perpetrated through systems, structure and organisation. The culpability for those past abuses rests as equally on the followers of those particular leaders as it does upon the historically demonised figurehead.

Personal Ignorance vs Social Ignorance
The ancient mystical sects of the Gnostics were great psychologists and one of the terms they employed to denote spiritual ignorance — Archon — serves as a wonderful vehicle for correlating personal ignorance to social ignorance.

In our habitual mind trapped in Self, we divide the world into Self and Other and that Self and Other into Superior/Inferior. Moreover, we consider ourselves "on top" and others underneath us:

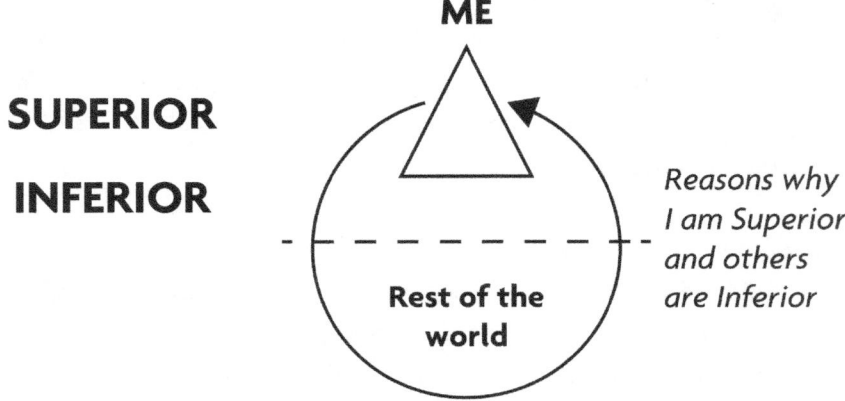

ME

SUPERIOR

INFERIOR

Reasons why I am Superior and others are Inferior

Rest of the world

A further understanding can be taken from the Gnostic Archon; human existence is social and is always relative to social systems, structure and organisation. The modern human mind perceives and builds hierarchies and its existence takes place within such organisation. For a fuller understanding of ignorance we must therefore move our discussion onto societies as a whole.

Power

Heirarchy

"Power corrupts. Absolute power corrupts absolutely."

I for one don't follow Acton's maxim; it all depends on your definition of power. In the journey from babe to adult we all grow in knowledge and understanding and develop greater scope of action and decision making, but we do not necessarily become more corrupt as this process proceeds. An adjustment to the axiom provides greater illumination:

"Control corrupts. Absolute control corrupts absolutely."

Human beings have the free will potential to act with or against the interests of others, to act creatively or to act destructively.

Our past experience is one of placing a single person, or small group of people, on top of a command and control hierarchy and then investing that person or group with supreme authority.

An arrangement that has been the recipe for human history.

Leaders and Followers
There is nothing inherently wrong with hierarchical organisation; in an organised cooperative venture there must be those directing and those following orders. The philosopher Bertrand Russell identified these social roles as positions of explicit and implicit power.

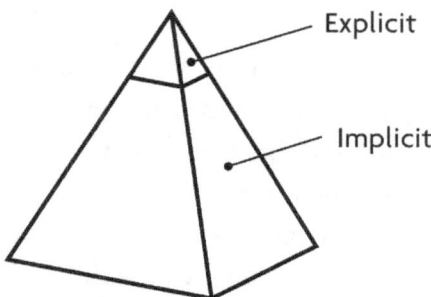

However, in a command and control structure there exists always the potential for extreme differences of interest between those in positions of explicit power and those in positions of implicit power. These potential differences of interest are enhanced by the fact that those with explicit power have greater capacity to design the systems, structure and organisation and thus influence those with implicit power accordingly.

Revolution

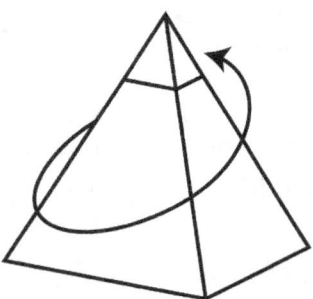

Acceptance of this existential scheme lies behind the historic assumption that social change can only take place through the seizure of a command and control hierarchy. The writer George Orwell reflected on this revolutionary pattern in human history and declared:

"Hope Lies in the Proles"

Hope does not lie in the Proles — hope lies in us recognising that there are neither Proles, a Middle nor a High.

Force of Empire

Social groups in themselves can be considered analogous to individual organisms. They must have internal capacities to maintain organisation, they seek to meet the needs of all internal components through growth and development, and they defend their own boundaries.

Problems can arise when growth creates clashes of boundaries or differences of interest between separate social centres. The endemic nature of conflicts arising from such can be witnessed by the imperial history of the world.

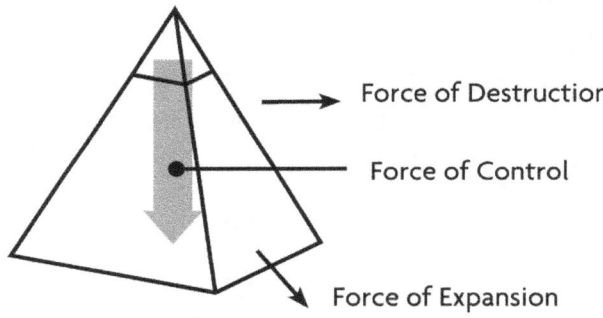

Corruption

When social organisation becomes corrupted through the abuse of power it becomes not a symbiotic relationship between those with explicit power and those with implicit power, but rather a parasitic one between those in a position of control and those in a state of dependency.

As within societies, so between societies.

This existential scheme can be seen in such examples as:

- historic slavery
- historic serfdom
- historic and concurrent landlordship
- historic and concurrent colonialism and imperialism
- our modern global financial system

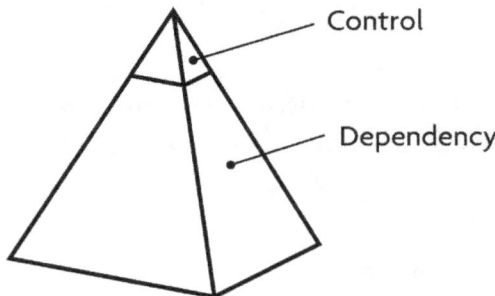

It is a motif that can be seen to varying degrees throughout many of our social systems, a most notable current example being the drive to gain greater corporate control of the profits from global food production through genetic engineering and the patenting of life forms.

"If a blind man follows a blind man, both of them will fall into a hole"

– Jesus

The Mind and Social Change

The archetypes of mysticism and self-realisation provide an excellent metaphor for understanding the solutions to our current global problems.

Ignorance & Separation
In the mystical traditions of the world, ignorance of the interdependent nature of reality arises from the primal separation we experience between the world "in here" and the world "out there".

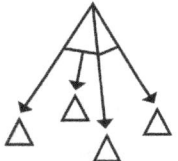

Self

Dependence	Control Relationships	Command
Single Benefit	Competition	Win / Lose

The solution to ignorance and separation is awareness and communication. Awareness is an experiential pathway that does not exist in theory or rationality alone.

Awareness (knowledge + conceptual understanding + life experience) + Communication

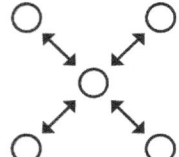

Non-self

Interdependence	Reciprocal Relationships	Communication
Mutual Benefit	Cooperation	Win / Win

Awareness + Communication = Creativity + Empowerment

Awareness and communication is the vehicle that gives new birth to the original meaning and purpose in our lives.

The Mind and Self-Defence

"It is easy to see the faults of others; we winnow them like chaff. It is hard to see our own; we hide them as a gambler hides a losing draw."

– Siddhartha Gautama

The main barrier to external social change is a personal internal one. Our mind protects its conceptual structure and self identity in the same way that it protects the body from an external challenge; through a fight or flight response.

Tower of Self / Tower of Not-Self   Ideas for Change

Belief System Reflex

Ideas for change often meet a spontaneous defence reaction based on existing beliefs. This reaction can be compared to the strike of a cobra — not only is it an automatic defence response but it can be dangerous; ultimately beliefs can kill.

While such responses may be illogical, the belief system reflex in itself is perfectly natural as minds must always protect their own integrity.

PART III:
Discussion

Validation

"He who knows the enemy and himself
Will never in a hundred battles be at risk"
– Sun Tzu[1]

It is not in my personal philosophy to have enemies, however, there are people whose plans for this world are the diametrical opposite of what I would like to see happening. Unfortunately for me, these are the people who are endeavouring to rule the thing.

The Grand Chessboard

A revealing glimpse into the minds of those currently in positions of global political power can be found in Zbigniew Brzezinski's 1997 book *The Grand Chessboard*. Within, he outlines what he believes are the geostrategic imperatives necessary to ensure U.S. global hegemony for the duration of the 21st century. Before discussing the central themes of this work, there are some notable socio-political observations contained therein which perhaps cast recent world affairs in a different light:

> "It is also a fact that America is too democratic at home to be autocratic abroad. This limits the use of America's power, especially its capacity for military intimidation."[2]
>
> "But the pursuit of power is not a goal that commands popular passion, except in conditions of a sudden threat or challenge to the public's sense of domestic well-being. The economic self-denial (that is, defense spending) and the human sacrifice (casualties even among professional soldiers) required in the effort are uncongenial to democratic instincts."[3]
>
> "The public supported America's engagement in World War II largely because of the shock effect of the Japanese attack on Pearl Harbor."[4]
>
> His Central Asian strategy to dominate 'The Grand Chessboard' is therefore unlikely to find approval "except in the circumstances of a truly massive and widely perceived direct external threat."[5]

To digress; 1997 was obviously a fertile year for geostrategic planning because it was also at this time that the neoconservative Project for the New American Century (PNAC) was founded. Containing the usual mixture of high profile academia and politicians common to Washington think-tanks, and featuring more than a few members of the Trilateral Commission, PNAC also promoted an aggressive strategy for U.S. global hegemony. They were also under no illusions as to the difficulty of promoting their own agenda, as made clear in Section V of their 2000 paper *Rebuilding America's Defenses*:

> "Further, the process of transformation, even if it brings revolutionary change, is likely to be a long one, absent some catastrophic and catalysing event — like a new Pearl Harbor"[6]

PNAC's goal for regime change in Iraq was clearly stated in a 1998 open letter to then President William J. Clinton which demanded "a full complement of diplomatic, political and military efforts"[7] in order to remove Saddam Hussein from power.

It's not within the scope of this work to ascertain if on that fateful day we were witnessing a 'False Flag' attack but, through whatever combination of circumstances, geostrategic dreams came true and a new Pearl Harbor arrived on September 11, 2001. By this time, many of PNACs members, including Richard B. Cheney and Donald Rumsfeld, were (re)occupying executive positions within the U.S. government. Events that day provided them an opportunity to turn plans into policy.

To return to Brzezinski's *Chessboard*; the thesis outlines the global geopolitical situation at the end of the 20th century, and identifies the supercontinent of Eurasia as the board on which the great game for global hegemony in the 21st will take place. Identified as the most crucial area is the 'New Balkans' of Central Asia; the energy rich and ethno-religiously volatile region where the interests of the great powers intersect. Perhaps the most salient points relating to ongoing international developments are the observations:

> "It follows that America's primary interest is to help ensure that no single global power comes to control this geopolitical space [South

West Asia] and that the global community has unhindered financial and economic access to it."[8]

"If the southern region [South West Asia] is not subjected to domination by a single player...American can then be said to prevail."[9]

"A possible challenge to American primacy from Islamic fundamentalism could be part of the problem in this unstable region. By exploiting religious hostility to the American way of life and taking advantage of the Arab-Israeli conflict, Islamic fundamentalism could undermine several pro-Western Middle Eastern governments and eventually jeopardise American regional interests, especially in the Persian Gulf. However, without political cohesion and in the absence of a single genuinely powerful Islamic state, a challenge from Islamic fundamentalism would lack a geopolitical core and would thus be more likely to express itself through diffuse violence."[10]

As an aside, what media label is applied to violence from such sources?

Violence that is expected in response to the mandates of U.S. global primacy.

In addition to its surface qualities as an accurate and informed overview of global political affairs, I find the greatest value of the book in its unspoken assumptions; assumptions which illuminate the political power dynamics of our world:

> All of the major factions on earth are locked in competition to find, control and then put into the air all of the fossil energy sources in existence.
>
> The environment, the source of life which makes all of these intrigues possible, is invisible to their military considerations.
>
> The only relationships between factions are Dominance/Submission. Win or Lose. If you do not use your power to dominate then it is assumed that you will be forced to submit to another.
>
> The emotional underpinnings of the whole endeavour are:
>> o Fear and Greed

All in all, this is a book I highly recommend to anyone who wishes to come to understand world affairs; for a book written in the last century perhaps the most marvellous aspect is the author's future predicting ability. Of course, Brzezinski has no crystal ball; his work is so accurate for two reasons:

- He possesses an excellent understanding of human nature and the dynamics of political power.
- The circles in which he moves are in a very good position to make the future.

Despite history unfolding in many areas as anticipated, not everything ventured in the book has come to pass, indeed the grand Central Asian ploy that the book is centred upon became largely obsolete once it became clear that the amount of oil under the Caspian Basin was much less than the estimates the strategy was based around.[11] This discovery would have shifted the axis of geostrategic importance further south towards Iraq and Iran.

As with all notable historic political power treatises the amorality of the analysis serves to camouflage the naked immorality necessary to achieve some of the suggested objectives, objectives of course dressed in the finest of language, for the noblest of purposes; considered justified for the greater good. Please note that by highlighting this work I am not wishing to demonise Brzezinski or the U.S. in particular (why cut off the head of the hydra when you yourself compose its body?); this book in fact provides a great service to humanity — an excellent insight into the competitive power games played by all nations and factions of the world. Games which do not consider the state of "The Prize" once the dust settles. It is these games of empire which must be tamed if we are to move forward.

In summary, if you are serious about solutions to our global environmental challenges then a useful assignment would be to educate yourself on the principles of sustainable development, then to read up on the plans of the globe's most powerful interests. Compare and contrast. On completion of this exercise congratulate yourself — because you will finally have arrived on Planet Earth.

It's not about using less water when you boil the kettle.

Unlimited Energy

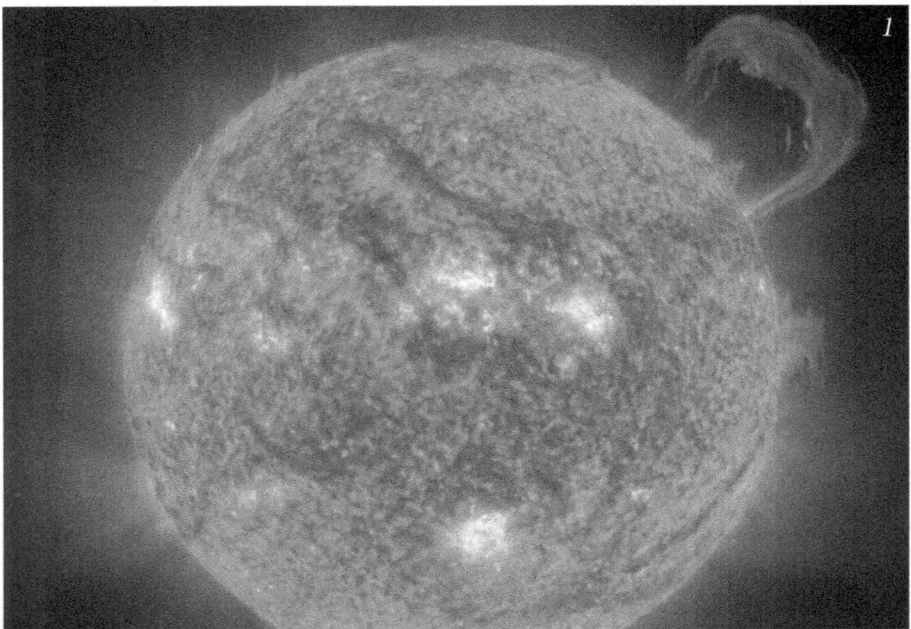

While all of the wailing and gnashing of teeth over energy depletion has been growing steadily louder over the last two decades, the planet itself has been continuing to circle around and around a certain great ball of light in the sky. From our friendly neighbourhood star, 1000 watts of power per square meter arrive on the surface of earth[2], enough energy to power the entirety of the living systems on the surface and the global differential flow of wind and waves. Dig down into the earth itself and the temperature increases by one Fahrenheit degree every 70 feet as one approaches closer to its molten mantle[3]. The flow of this liquid mantle around the core of the planet itself generates a gigantic magnetic field strong enough to protect life on earth from the extremes of the solar particle flow[4].

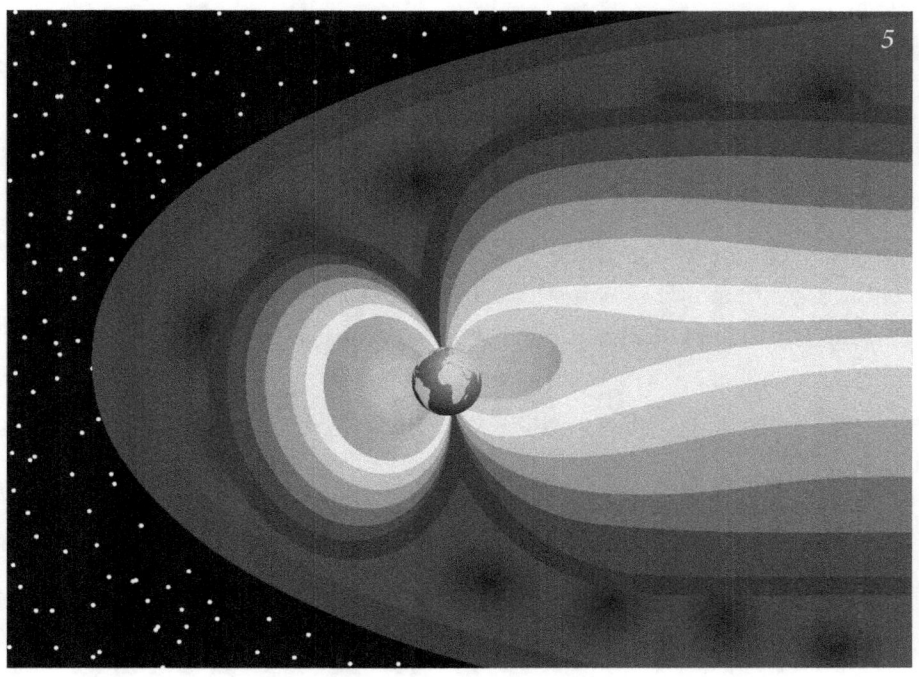

With this picture in mind, I'm sure we will be able to find some energy from somewhere.

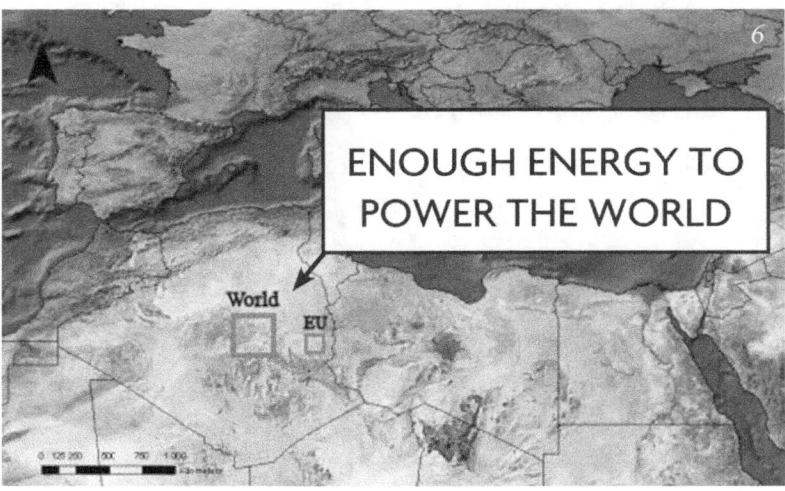

When considering the problem of energy supply, a useful mental image is to imagine modern industrial society as a technological organism whose food is energy sources; primarily fossil fuels, and whose blood is electricity. This technological organism runs parallel and is intertwined with the biology of the planet and the physical needs of humanity.

The good news is that the 'blood' in our analogy is limitless, and simply requires a differential of some kind for generation; the bad news is that the current 'food' supply which generates the blood has physical limitations[7]. The long term challenge then, is to *wean* this colossal organism from a fossil fuel diet, to an alternative energy supply.

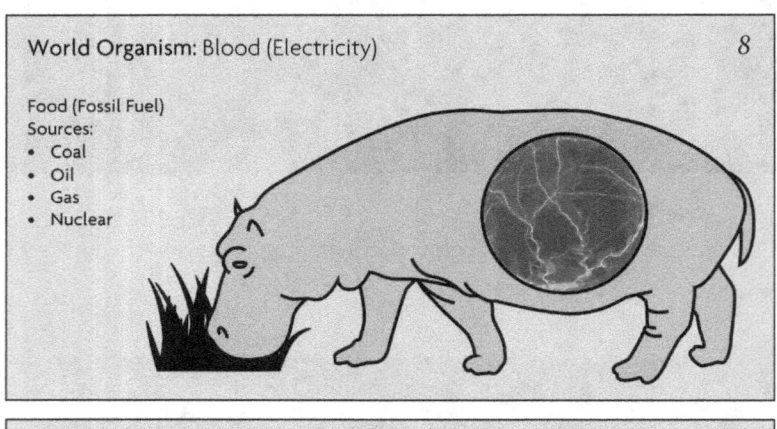

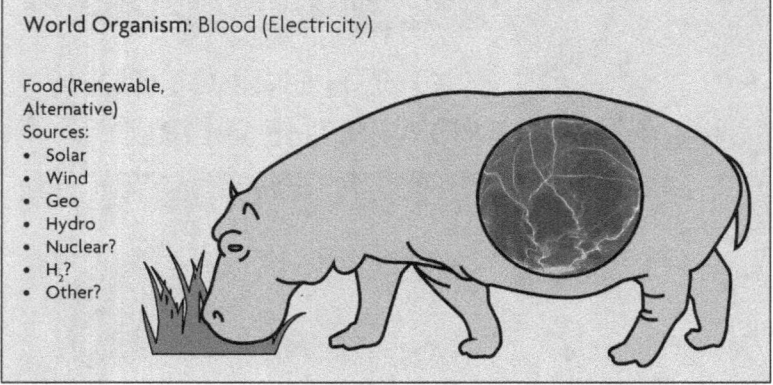

Of course, the issue of fossil fuel limitation goes hand in hand with the problems associated with excess CO_2 in the atmosphere. To bring about an energy transition and to solve these twin threats, fossil fuel use must be reduced simultaneously with alternatives being brought online.[9]

The diagrams above may provide some inspiration as to the availability of unlimited energy from the sun and the ability of renewable sources to provide for our needs, but the major hurdle to be cleared involves the current lack of a straight like-for-like alternative — there are important differences between fossil fuels and existing alternatives:

- Fossil fuels are a concentrated energy resource, easily portable, storable and possess the best and most scalable power-to-weight ratio of existing fuel sources[10]

To make use of renewable sources economic will therefore require breakthroughs in the storage and distribution of alternative energy supply. The good news is that technology in this area continues to progress, indeed a transitioned society could have orders of magnitude of more energy available than we do currently.[11]

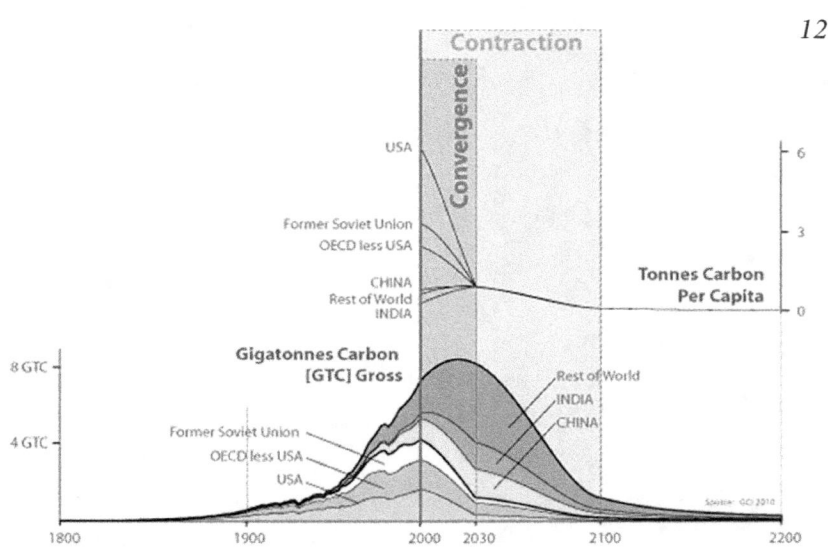

This example shows regionally negotiated rates of C&C.
It is for a 450ppmv Contraction Budget, with Convergence by 2030.

Approaches to how global net reductions in fossil fuel use could take place have been in discussion for some time. Those generally considered to be fairest are 'Contract and Converge' schemes, which involve nations agreeing to cap their CO_2 emissions then contract them down towards a pre-agreed global net target. The target is a global per capita (per person) standard, to which all nations would then converge. Those more developed nations with high emissions would have to drastically reduce their emissions, whereas those less developed countries would be able to increase their emissions for a number of years. A carbon credit trading mechanism would facilitate the process, with the more developed nations being able to purchase carbon credits and thus transfer resources to the less developed nations. In this way global net reductions could be hypothetically achieved whilst simultaneously promoting greater global equity, an elegant solution in response to the twin threats of global warming and fossil fuel depletion.

Contemplating Paradox, Part I: Personal

"do less, use less"

Perhaps by this point you may have noticed the irony of ironies of this book; the carbon footprint of my voyage. It is the new *original sin* and what you are reading is surely a climate change book written by the world's most evil man.

Well, if I was taking a moral lesson from my journey, it would be the very opposite of that which a modern day ascete would suggest. A discussion of which provides the perfect vehicle to explain why the environmental movement is so notable for its lack of success in catalysing the very energy transition that it so vocally calls for.

The carbon footprint of a book such as this stands in stark contrast to the "negative environmental morality" that you surely recognise. The perception that the use of energy and resources is "bad". In its championing of "do less, use less", the Green movement has created what amounts to the new original sin and the current rush of businesses to offset their emissions more than smacks of indulgences. The problem of excess carbon dioxide in the atmosphere is not sin, it is ignorance. Collective ignorance. The problem of excess carbon dioxide in the atmosphere cannot be solved through personal redemption, but only by collective action.

There is no "do less, use less" ethic in this book because if you hadn't noticed, this is a suicide ethic. The "Save the Planet" school of thought, or as I like to term it, the "Save the Money" movement. Because that is all that doing less and using less will do; save you money. And money saved will eventually be spent, it will flow into the world economy and it will turn the wheels of life, wheels which of course release greenhouse gases.

The wheels of life must always turn; it is the wheels that must go green, not the people.

Trying to solve climate change by using less energy is like trying to affect global money supply by using less money. Even if you were masochistically so inclined and did succeed in reducing your "money footprint", it would not stop other people from continuing to earn. This is more than an analogy because energy *is* money. *Black gold.* Due to its structural importance, and as a standard and store of value, fossil energy is actually much closer to being money in real terms than any of our *fiat* mediums of exchange. And we can understand human use of energy by our own attitude to money, which is illustrated in this simple equation:

$$Energy = Benefit$$

All the economically available energy in a living, growing system will be used and if this were not the case then we would commonly walk along streets paved with gold. If we begin from this starting point we would then realise that the only way to reduce the energy use in a living system would be to contract the input. The moral overlay of the Green movement very successfully obfuscates this simple observation. The concept of carbon footprints is found to be, if you will excuse the pun, no more than a smokescreen.

I understand the negative environmental morality because I found myself occupying this position as I immersed myself in the literature and searched for control of large scale social and environmental issues. I began to see my own decision making in increasing tones of morality, and consequently to view the decision making of everyone else around me according to the same criteria.

If you take this approach to social problem solving then you will draw a tight circle around what you can influence — your life — in an attempt to control what you are not in control of — large scale social and environmental conditions — and you will contract the circle of your life in reaction to any abstract information that you perceive as a threat to your continued existence. You will then become highly frustrated as you observe other people not conforming to the same limitations you have placed on yourself.

The negative environmental morality is simply the human need to control the world played out on the smallest of scales.

Cardinal offender of the new Green religion is one George Monbiot. *Heat* is an important book that I would recommend to everyone as it discusses the feasibility of a 90 per cent cut in carbon emissions by 2030, which is the magnitude and timescale of the changes to be made if we wish to attempt to avoid the worst predicted outcomes. Monbiot, like all observers who take climate change seriously, correctly identifies the problem as a systems issue and argues convincingly throughout his book why this is the case. However, despite isolating the presumably non-personal nature of the problem, he still cannot resist the temptation to pronounce moral judgement:

"Our moral dissonance about flying reminds me of something a Buddhist once told me when I questioned his purchase of unethical products. 'It doesn't matter what you do, as long as you do it with love'. I am sure he knew as well as I did that our state of mind makes no difference either to the exploitation of workers or the composition of the atmosphere. Thinking like ethical people, dressing like ethical people, decorating our homes like ethical people makes not a damn of difference unless we also behave like ethical people. When it comes to flying, there seems to be no connection between intention and action."[1]

First of all, an aside: "ethical living"; to even participate in this discussion assumes life in a certain culture, in a certain part of the world, to a certain level of education, with a certain degree of financial means and a certain access to automated transport networks. In sum, a certain standard of living and a certain energy content of the person and society in question. This is a luxury discourse and the ethical living movement is luxury decision making.

An example: Consider a shopper contemplating the ongoing extinction of fisheries. Faced with such an overwhelming large scale problem, it is easier for the individual shopper to go into denial, gain a perceived sense of control over such an abstract and become a fish moralist than it is for he/she to actually do something about the problem. This is forgetting that fish will still be fish, fish will still be eaten, fishers will still fish for fish, fish hauliers and fish sellers will still haul and sell fish, and economists and politicians will still oversee the overall begetting of fish. It is in absolutely everyone's interest — from fish eaters to fish to politicians — that the fishery be managed. However this has historically proven difficult, and not just because of the so-called "tragedy of the commons"; increasingly since the 17th Century the participants of the commons have had to contend with the black holes of compound debt and compound interest as a backdrop to their economic existence.

If enough people within society can recognise the problem and then act together in unison then it can be solved.

Contrary to the commonly intellectualised portrayal of society, there are in fact no evil consumers destroying the world; there are only people going to the shops. To label human beings as "consumers" is in itself dehumanising. The process of going to the shops is basically similar in all areas of the world with geographic and financial criteria having the biggest effects on decision making. If the everyday action of going to the shops imperils the future of humanity and the planet then this is a systems issue for society as a whole. As an individual you must continue to eat, meet needs and participate in the economy in order for both yourself and the economy as a whole to be sustained.

Indeed the environmental and related social ethical movements have provided the perfect camouflage for problems which can only be solved through changes to systems, structure and organisation. As an individual within society it is only possible to live inside systems, structure and organisation.

To return to the excerpt from *Heat*; Monbiots' opinion on air travel comes, without any sense of irony, in a work throughout which he waxes lyrical on his formative experiences all around the globe. Needless to say he did not have the same opinion on flight when he was working in Brazil, Ethiopia, Kenya, Tanzania or Indonesia (he did not get to these places by cycling). From this we are left with the impression that global human morality must be dependent on however George Monbiot happens to be feeling on any given day.

According to his own criteria and taking into consideration his biographic career activities and the amount of economic activity generated from the sale of his books and syndicated columns, the majority of human beings do not "destroy other people's lives" to the degree that George Monbiot does, and certainly not the average U.K. citizen whom he continually lectures on the inherent immorality of their existence.

Anyone employed in a position where fossil fuel dependency is more obviously part of daily life than that of a journalist — such as an airline pilot, a coal miner, an offshore oil worker or a farmer — would correctly rationalise that if they did not do their job then someone else would take their place and the situation would not change. They would correctly rationalise that if they did not do their job then whatever other means of employment they took on would still be dependent on the existence of the former.

So much for "Green Ethics".

It is nothing but the egotism of the self-righteous.

Now, if we are looking at all of George Monbiots' positive contributions to society; his original thought, passion for change and career of activism, we find them to be a direct example of:

Awareness (knowledge + conceptual understanding + life experience) – Communication

= Creativity + Empowerment

What separates someone active and aware from the average citizen is simply **life experience**. It does not come from anywhere else, it cannot come from anywhere else.

My ire with the "do less, use less" movement is not just because it mistakes money saving for action but because in the final analysis it is a statement of the *status quo*. It suggests that the problems facing us can be solved if only we do a little less, use a little less, tighten our belts, be a little more *ethical*. This is an entirely false position and to promote this position is to promote falsehoods.

The average person, on becoming concerned by the encroaching environmental mega issues may endeavour to find out what he or she can do only to be confronted by an endless list of "do less, use less" choices which amount to, at best, student money saving tips and, at worst, various elaborate methods of self harm. Turning your heating down by one degree for example does not "Save the World"; it just makes you cold.

If you embrace this approach, then the drive to work every morning becomes a statement of global annihilation rather than participation in the cooperative venture supporting civilisation. This very effectively strips all meaning and purpose from existence. Human beings are all emotionally wired not to do harm and the Green association of daily existence with harm can only result in serious psychological trauma.

The practically-minded on hearing the "do less, use less" message, rather than dwelling in guilt and abstractions, will continue on with life regardless and increasingly perceive environmentalism as something 'not-self'. This will result in an increasing hostility to ecological awareness and produce the net result of compounding social ignorance.

The abstract-minded *will try* and "do less, use less" and thus become increasingly inactive. Such a constant state of guilt can only lead to existential and intellectual paralysis. This explains why the most feted Green literature is full of elementary logical errors, especially in the economic sphere. The Green movement lacks experience in commerce and industry because enterprise and innovation are based on creativity and reward and not on self-sacrifice and self-denial.

Indeed, the sanctimonious religious nature of Green extremism is due to its basis in unrecognised psychological conflict. It is not possible to "do less" and also "earn money" at the same time. On the topic of Greens making money there seems to be no connection between intention and action.

Cognitive dissonance indeed.

The personalisation of environmental issues is the path of extremism. There has never been a personal environmental issue and there never will be. At the farthest reaches of this kind of extremism is the position in which your ego becomes the natural world and other human beings correspondingly become the enemy. Furthermore, if you try and "ethically" contract your existence in response to abstracts then the logical conclusion of this will be suicide as your only moral option.

This narrow focus on the individual entirely forgets the fact that all of the institutions, structures, systems and organisation of society exist, and the very reason for their existence is in order to manage problems greater than the individual.

The Green movement can be seen as an example of the usual features of human dualistic perception; separating the world into those who buy energy saving lightbulbs and those who do not, a binary morality where the use of energy and resources is "bad" (unless it's you that's doing it), and a complete obsession with symptoms, be it SUVs or aircraft, at the expense of the disease of total fossil fuel dependence. If mainstream society serves as an example of ignorance of ecological interdependence, the Green movement can correspondingly be caricatured as ignorance of economic interdependence.

The only way to actually "do less, use less" is to "earn less, spend less" and if you were to take this personal course of action you will be in serious trouble within a short period of time. When workers compete to earn the least wages and compete to not get promoted, when new graduates compete to find the least pleasant and least paid employment and when students compete to finish bottom of the class, then it can be finally said that human beings have gone "Green".

This may be *reductio ad absurdum*, but inference is important in issues which are processed subconsciously and emotionally.

I find it especially troubling that this mindset is being fostered on the younger generations of the world. If turning off the lights and using less energy in the household is "moral", then the entire remainder of that child's existence becomes a statement of immorality. Under this logic, rather than getting an education with the aim of best possible future participation in the economy, the most moral course of action for any young person is to become a heroin addict — because then they will not do anything, use anything or go anywhere.

Mass green morality is mass de-moralisation.

Being is acting and the extremes of the environmental movement with anti-energy and anti-action philosophies arrive very effectively at an anti-life position.

To return to solutions; we have already identified that only a systemic measure can contract the net energy use of a society. However, in practice what this would entail is increasing the scarcity of fossil fuels and correspondingly increasing their price. This creates a number of problems, the most important of which are to do with food production. Fossil energy is integral to modern food production due to: the use of fossil-source fertilisers, the energy required for irrigation, the energy required to manufacture farm machinery, the energy required to power farm machinery and, not least, the energy to power the transport networks necessary to distribute everything, including food. Less fossil fuel equals less food.

In modern society fossil energy is a fundamental; both for production of materials and for operation of transport networks. Increasing the scarcity and thereby increasing the price of fossil fuels creates ripple effects across the entirety of economic life which can be summarised as; *negative economic effects*. Now, *negative economic effects* into *positive interest money* systems simply **do not go**. Interest based monetary systems cannot 'Contract and Converge', they can only 'Contract and Collapse'. Any modern nation which attempts to contract its fossil fuel use in the absence of a simultaneous transition to at least a like-for-like alternative would discover what people increasingly since the 17th century have done in times of economic crisis — that money does not exist. The build up of negative economic effects would result in people losing their jobs and consequently withdrawing their savings, runs on banks would result and the entire commercial banking system, now global in nature, would be vulnerable to a complete collapse. Historically, such a situation has benefitted only those sectors of society who have financial reserves and are in a position to restart the credit cycle anew once the collapse is over.

The leads to the understanding that in the absence of any systemic methods to reduce fossil fuel use, 'Contraction and Convergence' or otherwise, the cyclic nature of a debt-driven global economic system together with the limited nature of the resource itself mean that such a situation is only a matter of time.

This is a crushing realisation.

It serves to underline our complete dependency, and our vulnerability in the absence of an available like-for-like energy source. Without a simultaneous energy transition, net fossil fuel reductions would have more catastrophic effects on society than any of the outcomes witnessed under IMF austerity plans.

No Energy = No Benefit

In conclusion, if you come to understand the differences in real world conditions between North Korea and South Korea and how they came to be, then you will also come to understand why I consider promoting an anti-carbon ethic to be completely immoral.

Contemplating Paradox, Part II: Social

"do more, use more"

Given that argument over systems, structure and organisation divided the earth during the last century, and the glacial pace of global institutional change at the best of times, it is difficult to foresee the required response in the optimum time. Faced with such a devastating mental scenario, it is best to take the advice of the wise and turn our attention away.

This brings us back to the oft asked question — **what can one person do?**

To answer this I take inspiration from the philogenic roots of ecology:

Eco Logos

Keeping your house in order.

To illustrate this answer this I draw inspiration from those 'one persons' who are taking action and solving problems, at both the grassroots and institutional levels, and that answer is to:

"do more, use more"

Take the example of many of the properties of the WWOOF movement; where individuals, families and small groups of families have gone a long way to show what could be done *right now* to move towards a complete energy transition and the sustainable management of land. Note that all of these achievements have been made without the use of any new technology, and without the focus, power and creativity of mainstream society. Furthermore, these changes herald the potential for an increased quality of life for the developed world, and an increased material standard of living for the developing world. They represent the opportunity for an enhanced, not a diminished, human existence.

Looking at the example of my own life, if I want to have greater personal practical involvement in any of the emerging fields I am interested in; from renewable and alternative energy, sustainable agriculture & land use, land regeneration, ecodesign and grassroots democracy to complementary currencies, then I will obviously have to:

"do more, use more"

Keeping your house in order can be broken down into two levels of approach:

– Personal Practical (Own House)
– Socio-Political (The Social House)

Taking my own home into consideration, the efficacy of solar power is rather less in Scotland than in Australia and the price of modification and conversion to geothermal power prohibitive at time of writing. Indeed, most people spend the majority of their life paying off the interest-bearing loan on their property, so to expect them to convert their houses to be powered exclusively by renewable energy is not currently a feasible option.

However, rather than spending large amounts of money to convert my own home, a move which will benefit mainly myself and my family, if I instead use my socio-political impetus to call for my society to utilise the potential wind and wave power off its coastline, this is an action at little to no personal cost which will potentially benefit not only myself and my family but *millions of other people at the same time.*

The socio-political approach therefore is clearly the most important response, and this is something everyone in modern society has, regardless of level of income and regardless of means of employment — off shore oil worker, farmer, airline pilot, coal miner and so on.

(Please note that the purpose of these examples is to outline the personal element of social problem solving. If I was giving out any personal advice to my reader I would say — *aim to live a full and complete life and endeavour to reach your highest potential*, because this is what I intend to do with mine.)

By championing the grassroots approach I must be careful not to promote another Green smokescreen; the lifestyle smokescreen. Again, such approaches are only likely to lead to disillusionment due to the prohibitive financial criteria involved. They are also not practically feasible either; if the entire population of Hong Kong was to decamp to the New Territories to live on solar powered organic farms, for example, this would cause far more environmental problems than it would solve.

There are no personal negative solutions, there are no lifestyle solutions; there are only social solutions. Positive social solutions. Issues larger than the individual, be it global warming or peak oil, can only be either psychologically accepted or practically faced through social goal setting and a sense of common purpose.

In contrast to its current overriding emphasis, environmentalism should be reframed as a call to people's highest ideals and a challenge to their highest potentials. It should be seen as positive social goalsetting for mutual benefit; the promotion of win-win exchanges between people and win-win exchanges between people and nature.

The moral overlay must be dropped. Environmental problems are not moral problems, they are practical problems. Practical problems which can only be solved by practical means. People are at heart creative and constructive and the necessity of developing our most heartfelt qualities in response to the challenges facing us should be made clear.

In contrast to its current overriding emphasis, environmentalism should express the importance and necessity of the personal use of energy and resources. It takes energy and resource to keep your house in order. It requires a positive attitude to energy and resources to fulfil the direct responsibilities of life and family. It requires a positive attitude to energy and resources to be able to play an active role in society.

This leads to an understanding of the importance and necessity of the sustainable use of the world's renewable resources, and the importance and necessity of developing cyclical uses of the world's non-renewable resources. The goals of environmentalism should be seen as what they are — life enhancing and life affirming — and such a perspective shift requires a greater eco-literacy to spread throughout society as a whole.

The aim of the preceding essay — Paradox I — was not to completely discount personal lifestyle changes, rather to ensure they are considered in their full context. If we look at changes that are a result of personal choices — growing interest in fair trade, sourced products from sustainable forests and fisheries, organic foods, complementary medicine, grassroots activism, and so on — we can once again see that these choices are an example of:

Awareness (knowledge + conceptual understanding + life experience) +
Communication

= Creativity + Empowerment

"The Recipe" for positive change can be identified once more. It does not come from anywhere else; it cannot come from anywhere else.

In contrast to the Green message, taking a personal practical involvement in any of the emerging fields is amongst the most expensive courses of action that any individual can ever take, and is very obviously not a case of:

"doing less, using less"

Likewise, to have a greater socio-political input into any of the emerging fields involved in sustainability — renewable and alternative energy, sustainable agriculture and land use, land regeneration, ecodesign, grassroots democracy, complementary currencies and so on — as with any other field of life, requires:

"doing more, using more"

It requires using transport networks *more often*.

To influence society at the local, regional, national and global levels requires correspondingly increasing magnitudes of energy and resources. This is:

"doing more, using more"

The emphasis of the environment movement must change from things that you don't do, to things that you do:

Managing the sea, fishing sustainably, foresting sustainably, reforestation, restoring ecology, revitalising agriculture, farming organically, managing wetlands and rivers, clearing up pollution, recycling, transitioning energy sources, researching alternatives, developing new technologies, being more active in the community; all these require action.

All these are _things that you do_.

All require a positive attitude to the use of energy and resources.

If you don't do them, then they are not going to happen.

Without individuals employing their personal energy and resources, they are not going to happen.

To summarise, if you are concerned with any of the issues discussed in this book, then I would suggest you inform yourself, utilising examples such as the bibliography given, as well as examples of opposing and differing viewpoints given in other literature before forming an opinion. Once formulated, if you do perceive that there are problems to be solved — then do something positive.

Reconnection

Faced with such a scenario of globally interconnected problems, it is clear that we are currently living in interesting times. But in every Crisis there is Opportunity, and globally interconnected problems herald the possibility of globally interconnected solutions. Solutions with the potential for transformative change for the betterment of both humanity and the environment.

The solution to the twin threats of global warming and total fossil fuel dependency is a global energy transition, and the solution to ongoing social and environmental degradation is a 'path of sustainability':

Global Energy Transition + Path of Sustainability

This "global energy transition + path of sustainability" will be a dynamic process, rather than any kind of static endpoint, which will change, develop and grow in the implementation, evolving as does nature.

Such changes to society have long been called for by those in the environmental community. However, even the language of the sustainability debate often reflects the old paradigm — a social drive towards an energy transition is often thought of as a "Manhattan Project"[1].

What was the Manhattan Project? A massive centralised attempt to gain total power over 'The Other'.

I prefer a different analogy; that of a Moon Mission. However even the, at first glance, comparatively neutral social goal of a "Moon Mission" was still a historic attempt to gain dominance over 'The Other' by achieving primacy over a whole new theatre of action.

In contrast this will be a movement with no 'Other'.

As with a 'Moon Mission', it will require general social goals to be set; goals where the exact *modus operandi* will not be in existence at the outset, where the details will need to be filled in along the way. There are sure to be both unforeseen hurdles and unforeseen breakthroughs.

But if the initial goal is not set, then no journey can ever take place.

And in contrast to the historic Moon Mission, this movement will begin from the bottom up and not the top down.

A "global energy transition + path of sustainability", in practice, is the decentralisation of energy, the decentralisation of money and the devolution of power.

Crucially neglected by current mainstream thinkers, the most important element in such a transformation is a <u>global monetary transition</u>. This global monetary transition is primary, and is necessary to enable the secondaries of global energy transition and path of sustainability to take place. Without a global monetary transition a "global energy transition + path of sustainability" cannot occur:

Global Monetary Transition -> Global Energy Transition + Path of Sustainability

$$1^{\circ} \qquad 2^{\circ} \qquad 2^{\circ}$$

Positive interest monetary systems catalyse competition and short-termism between economic actors.[2] New negative interest money systems at the community levels will enable greater personal empowerment, and will unleash the creativity and ability to solve problems that currently lies latent in society, masked by usury and competition.[3] This empowerment will enable the grass-roots economic and political changes necessary towards energy transition and the sustainable use of resources.

The reformation of international institutions — the 'top down' approach — will be a secondary expression of the 'bottom-up' effect of the greater personal empowerment enabled by complementary currencies. The current best proposals for reformation of international trade are monetary visionary Bernard Lietaer's **Terra**[4], or environmental economist Richard Douthwaite's **EBCU**[5].

The **Terra** is a basket of commodities intended to be used as a global reserve currency, designed to counteract the business cycle and to encourage sustainability through a negative interest rate. Whereas the **EBCU** is a reprise of Keynes' original Bretton Woods proposal, with fossil energy and permits to emit greenhouse gases taking the place of the gold standard and US Dollar.

With either a zero or negative interest international currency, world trade can be made balanced and self-regulating. The **EBCU** proposal in particular would be an ideal 'weaning' mechanism for a global energy transition, though it would also require (i) the circulatory quickening effect of local complementary currencies, (ii) stable national banking, and (iii) a simultaneous aggressive global energy transition, in order for the global economy to continue to function with the background of net fossil fuel reductions taking place.

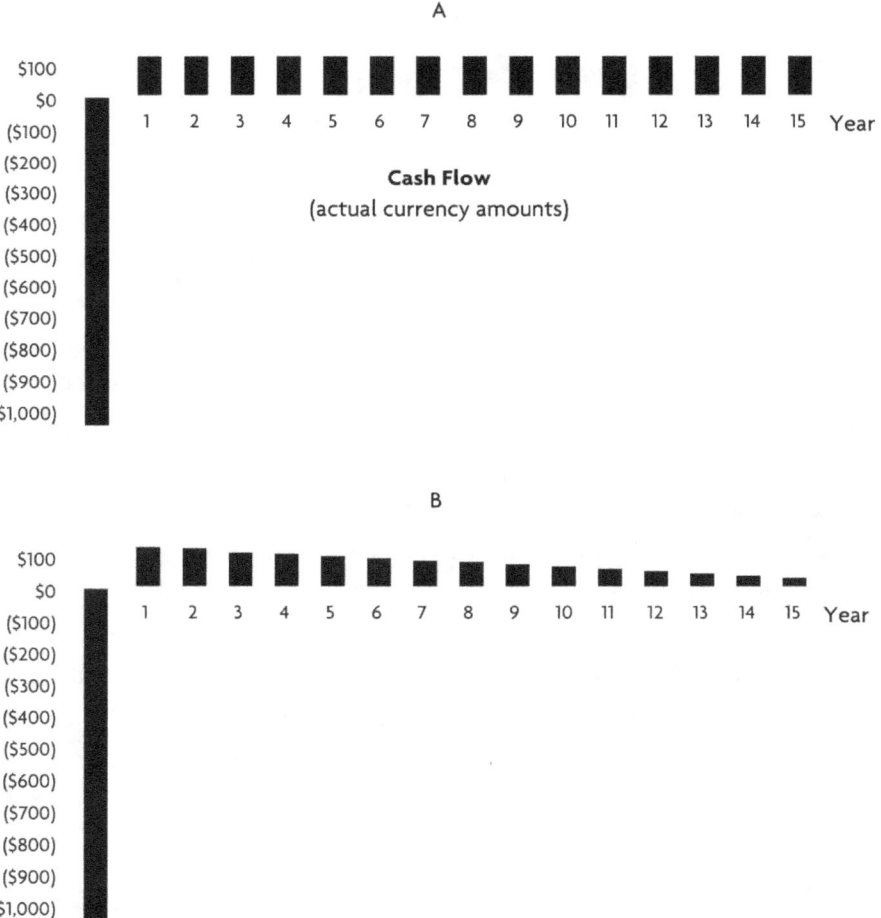

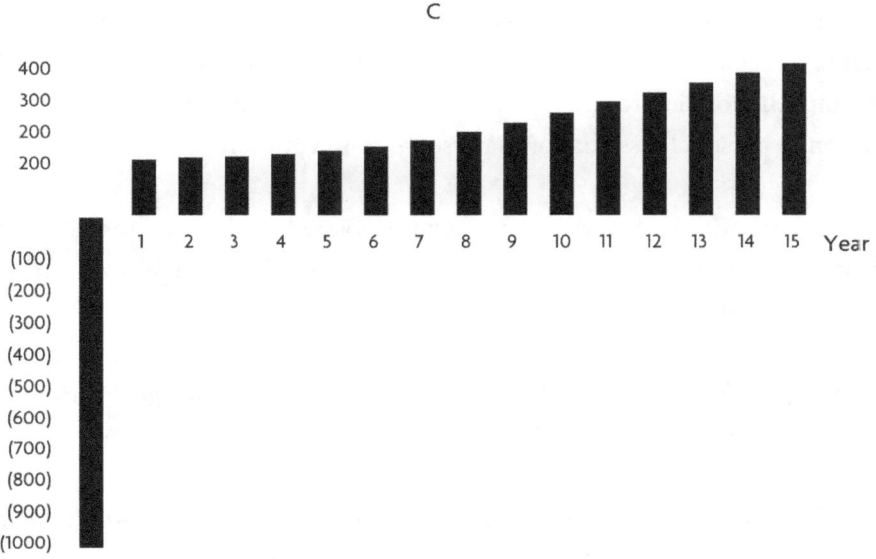

C

Fig A shows a project with a $1000 up-front investment yielding a net profit of $100 every year for fifteen years. Figure B shows the true picture of this cash flow when taking into account a 10% interest rate: the future is found to be less profitable than the present. Alternatively, Figure C shows what the cash flow of the project would be like with a negative interest rate: where the future is more profitable than the present!

Negative interest currencies catalyse cooperation and long-termism between economic actors.[7] Complementary currencies with a negative interest rate, as illustrated by the Miracle of Worgl, hold the potential to revitalise urban and rural communities *in every single part of the world*. This rebirth of society would enable the global energy transition to get underway. The reformation of global trade would allow a cooperative world order to emerge and enable societies and their public and private agencies to plan and act on long term lines. This freeing up of human potential would allow the restoration and regeneration of our planet, would reconnect us with ourselves and each other, and create a new flowering of literature, art, music, science, architecture, technology, community and spirituality: **culture**: on a global scale.

All of the great problems facing us today have their origins in disconnection, and their solutions lie on the other side: in reconnection. The solution to

economic collapse is economic revitalisation, the solution to ecological degradation is ecological regeneration, the solution to global asymmetry is global
harmony, the solution to social dislocation is reconnected community and the
solution to personal dissociation is personal re-integration.

Not only this, with such a transformation the possibility of an end to that great
terror of past millennia — large scale war and the causes of large scale war —
comes within sight.

The possibility of a world transformed awaits! Within a generation!
The opportunity is there. Will you take it?

*There is nothing more difficult to carry out, nor more doubtful of success, nor
more dangerous to handle, than to initiate a new order of things.*

The process of "Global Energy Transition + Path of Sustainability" will take
place through a:

- Decentralisation of money
- Decentralisation of energy
- Devolution of power

However, energy and money are the great levers of World Empire. Such
democratising liberation will not be a popular choice with those who profit
from the existing world order, whose power and means of opposition are
exemplified by trillion dollar profit lines and trillion dollar weapon systems.
Oligarchic Globalisation has been the exact opposite of a 'global energy transition + path of sustainability'; a process of the ongoing global centralisation of
political power in supranational bodies, the ongoing global monopolisation of
energy resources by multinational corporations and their military allies, and
the unquestioned and almost total supremacy of interest bearing fiat money
creation governed by supranational agencies as a global standard; all with the
consequent rise of a global monoculture.

For the reformer has enemies all who profit by the old order, and only lukewarm defenders in all those who would profit by the new order.

The 'do less, use less' Green approach began with well-intentioned activists aiming to promote a measure of personal impact into global problems. However, it has been seized upon and sculpted by those most special of special interests for its potential as a vehicle of global centralisation towards their aims of one world government:

Global problems require global solutions.

Furthermore, it's potential as a tool of entrapping the mass populace within a constrictive mental focus whilst simultaneously maintaining the existing psychology of scarcity on which usurious systems of exploitation depend has been utilised.

"Hegemony is as old as mankind"[8]
– Zbigniew Brzezinski

World Empire has long been the political aim of the ages. One world government is now in sight and the major factions make their moves in the fossil fuel endgame. To monopolise and extended the profit lines of perpetual resource transfer indefinitely into the future, whilst maintaining the mass populace in a state which allows such exploitation, is the goal.

So, for the potential of systemic solutions we have instead the *status quo*, with the planned addition of another layer of taxation on a global scale, placing even greater limits and controls on the people, all under the watchful eye of growing technological surveillance and backed by a creeping repeal of civil liberties, justified in fear of terrorists or whatever other manufactured adversaries are presented, engaged as we are in manufactured wars.

That's why the 'do less, use less' Green approach is noticeable in its worldwide ubiquity and "Global Solutions" which are knowingly unfeasible — to those who understand money creation, such as the aforementioned special interest groups — are prominently politically championed...

...and all the while solutions that work at the systemic level experience a mass media black out.[9]

If you don't know about it, then it doesn't exist.

If you don't believe in it, then it doesn't exist.

This lukewarmness arises partly from fear of their adversaries who have the law in their favor; and partly from the incredulity of mankind, who do not truly believe in anything new until they have had actual experience of it.

So there we are, bewildered by a constant stream of global problems apparently without genesis, projecting our responsibility onto leaders offering "Change" but serving the interests of the status quo, endangered by enemies without; nations abroad, suspicious of our neighbours, endangered by enemies within; inherently flawed, not trusting ourselves. Anxious for the soporific benefits of electronic toys and an addictive lifestyle, suckling on the false reality of a media that distorts and misinforms, a media that offers the catharsis of enemies destroyed.

Diverging the world into 'Us' and 'Them' and figuring out how to be on the side of the 'Winners'.

It is not about 'Us' or 'Them'. It is not black or white. It's not about Leaders or Followers, Right vs. Left or Capitalism vs. Socialism. It's not about environmentalists vs. multinational polluters. It's not about Christians and Muslims, West vs. East or North vs. South. It's not about the Pope, Ayatollah Khomeini or the Dalai Lama. It's not about Bush or Blair, Obama or Cameron. It's not about Saddam or Osama. It's not about H_2O, CFCs, HFCs, HCFCs, CH_4 or N_2O. It's definitely not about Carbon Dioxide.

It's about Consciousness.

Well, so think I.

PART IV:
Epilogue & Conclusions

Epilogue: An alternative look at 2007

After life on air for the best part of two years, those first few weeks were life in treacle. But the sense of normality and its rhythms soon returned and I settled back in. I was still me and family and friends were still family and friends. Yet everything was at the same time subtly different. You can really never go home again. My 'no longer go to church' stance, for example, caused some consternation in the first instance ...but they got over it. On TV David Attenborough was telling people to use less water when boiling the kettle.

I took in a game at Celtic Park and couldn't help but remember attending a baseball game in Hiroshima, where the families sitting around me in the stands gave me foam banners and encouraged me to join in; a real friendly, fun, uplifting and genuinely family atmosphere — where supporters actually supported. Instead, I was now inside a giant cauldron of negativity; Paul Telfer is a player of journeyman ability, but that does not make him a *'fucking wanker'*.

Selling my apartment at the peak of the property market had been the right decision, and the debt I had dug into in the latter part of my time in Australia and on my way home disappeared painlessly into the large profit I had made. A profit substantial enough to allow me to begin the writing and further researching of this book you are now reading...

February:

What with my recent experiences and the topic of my writing and research, I was most interested to see *The God Delusion* by Richard Dawkins ubiquitous in bookshops everywhere. It is a book that in fact does a great service for spirituality in its questioning of belief and rational lysis of dogma.

A wide ranging and well argued book, I'd recommend it to all believers, non-believers and agnostics. However, his central thesis that annihilationism raises consciousness can be falsified by any time spent in countries where institutional annihilationism was imposed. This resulted in a very obvious lowering of consciousness.

I also couldn't help but noticing that the route of personal experience was somewhat dismissed, deserving all of three pages.

"That is really all that needs to be said about personal 'experiences' of god, or other religious phenomena"[1]

Well Richard, I have some more to say on the route of personal experience; it is how we discover:

The Truth

Dawkins bases his dismissal on the fact that all perception is a relative conditioned model of reality. This is not a new philosophical idea, and how we interpret its meaning depends on our assumptions about an objective reality, and the link between outer reality and our subjectivity.

He also can't resist the temptation to heap scorn upon the route of personal experience, conflating it with the drunkard's hallucination of "pink elephants";

"well some people have experienced a pink elephant, but that doesn't impress you"[2]

He continues, conflating personal religious experience with serial killers, mass murderers and the seriously psychologically disturbed. So, if anyone has a personal experience that Richard Dawkins doesn't agree with or cannot prescribe meaning for, they can be heaped into this category? This is not rational discourse, this is a pogrom; a defamatory strawman argument.

Personally, I don't think that such epochal addresses as *The Sermon on the Mount* or *The Turning of the Wheel of Dharma* were inspired by communion with pink elephants, nor do I think that the sea changes in human consciousnesses and behaviour that resulted from such addresses were a result of identification with pink elephants. I don't think that the emotional and intuitive basis of religious belief is rooted in pink elephants, and I don't think those that sought and found direct experience; people from St Francis of Assisi to Meister

Eckhart, Guru Nanak to Adi Shankara, Milarepa to Hui Neng, Rumi to Hazrat Inayat Khan, Siddhartha Gautama to Lao Tzu were actually inspired by and directly experiencing the inner pink elephant.

Dismissing and diminishing the route of personal experience is not a scientific attitude.

Co-incidentally, a less entrenched external scientific perspective on the mysteries of life was provided by Paul Davies in *The Goldilocks Enigma*. A book that seeks to answer the question of why planet earth seems, like Goldilocks porridge, to be 'just right' for life.

It is a stimulating intellectual odyssey that I would also recommend to everyone. I think it is a book that ultimately raises more questions than answers. And I do take issue with the statement:

"Attempts to gain useful information about the world through magic, mysticism and secret mathematical codes mostly led nowhere"[3]

This tells us more about Paul Davies' values and perception than it does about mysticism. I am sure those historical persons listed above would agree with me. However, he does admit:

"I have made no attempt to consider other modes of discovery, such as mysticism, spiritual enlightenment or revelation through religious experience."[4]

It is my assertion that this is the only place where the answers to the "Big Questions" he is asking can be found.

March:

I went to church with the family. You can't really get away from Weddings, Funerals, Christenings and the like. That sensation of being dipped in a vat of acid once again.

In other news, the world slave trade is estimated to involve twenty seven million people, more than at any time in history.[5]

May:

The march of elite-driven supranational government continues apace with the expansion of the North American Free Trade Agreement (NAFTA) between the United States, Canada and Mexico into the Security and Prosperity Partnership (SSP).

Mooted to "increase security and enhance prosperity" amongst the three nations, in practice the SSP places Mexico and Canada in a subsidiary position to the United States.[6] The United States military borders will be pushed out around Mexico and Canada, whilst greater access to their resources will be acquired, all in the name of "freed trade" and "security".[7]

How much do the people or legislature of these countries know about this creeping loss of sovereignty? Would they have accepted these agreements if they did know? Would they have been able to reject them even if they had wanted to?

June:

Our very sick world economy continues to show signs of imminent market crash.[8]

The sanctity of my hometown, Glasgow, is violated by terrorists.

Two Muslim men attempt to drive an incendiary-rigged car through Glasgow airport at the height of the school holiday season. Thankfully hapless, they fail in their mission. Airport staff, police and passers by restrain them and put out the flames.[9] One of the attackers later dies from his burns, the other is tried and imprisoned.

The response from government and mainstream society is predictable; yet more evidence of the need for the "War on Terror". This chance for society as a whole to discuss the lives and actions of the violaters; to attempt to see their perspective of opposition, understand the genesis of their hate or to recognise their personal sickness and sense of powerlessness; is not taken. Instead they are "evil".[10]

If you trigger the death of over a million people[11] and the maiming and disenfranchisement of millions more[12,13,14] and remain oblivious to the fact that you have caused this death and mass torment, you can't expect this not to destabilise the sanity of those who see life from the perspective of the dead, maimed and dispossessed.

We are assailed by a handful of the desperate with homemade bombs, while we industrially and institutionally assail under the cloak of "legality" with combined national armies and trillion dollar weapon systems.

July:

Serious flooding hits the U.K.[15]

This summer is one of the wettest since records began.[16] It is generally accepted as a symptom of global warming, however the prescribed governmental cure very old paradigm; more investment in flood defences and nuclear power.[17]

Adaptation over mitigation.

Centralisation over an energy internet.

August:

Six U.S. nuclear warheads go missing.

They are eventually recovered 36 hours later, several hundred miles away at a different airbase, having been armed on advanced cruised missiles and flown cross country by a B-52 bomber.[18]

This complete breakdown of military protocol in the world's sole remaining hyperpower appears as a light relief story in the humorous section of tabloid newspapers a few months after the event.

Completely ignored by the world press are the deaths of several military personnel from the two airbases involved in the incident, in the weeks either side of the event, in a series of apparent accidents.[19]

September:

Opium production in Afghanistan for the year 2007 is projected to reach a record high, up 34% on 2006.

Helmand province alone is now the world's largest supplier of illicit drugs.

The president, Hamid Kharzai, is fiercely critical of the 'international community', claiming direct impediment to the counter-narcotic strategies of the Kabul government.[20]

November:

The FBI raid and seize the assets of the grassroots creators of the "Freedom Dollar"; a 100% asset-backed currency exchangeable for U.S. Dollars.[21] Neither the event nor the reasoning behind the alternative currency make it to the major national or international news platforms.

The IPCC release their Fourth Assessment Synthesis Report.[22]

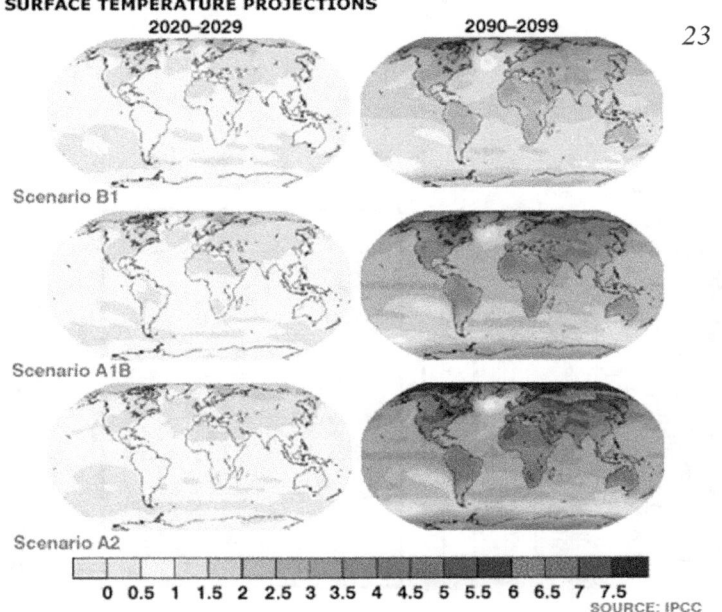

SURFACE TEMPERATURE PROJECTIONS
2020–2029   2090–2099   *23*

Scenario B1

Scenario A1B

Scenario A2

0 0.5 1 1.5 2 2.5 3 3.5 4 4.5 5 5.5 6 6.5 7 7.5
SOURCE: IPCC

1990   **2007**

What differences are there in the predictions between this new report and the original?

1990   *24*

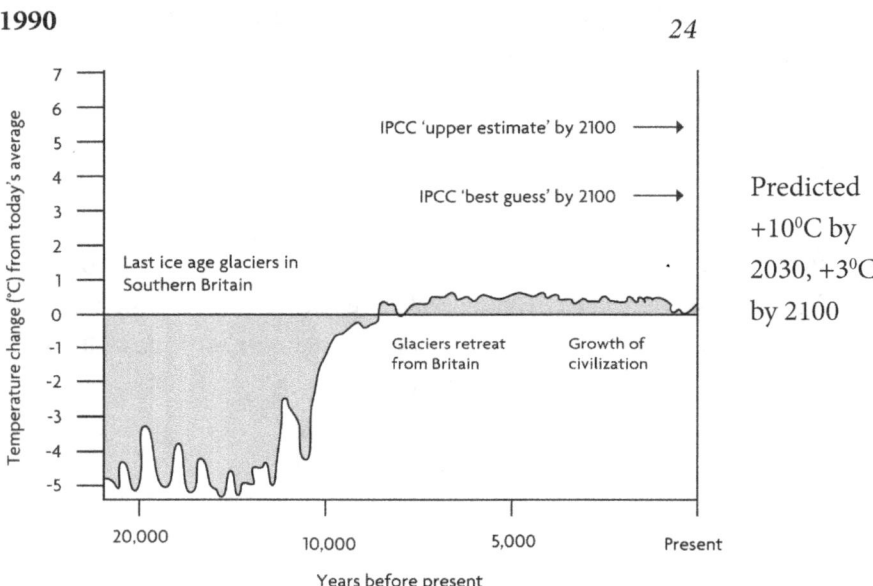

Predicted +10^{0}C by 2030, +3^{0}C by 2100

2007

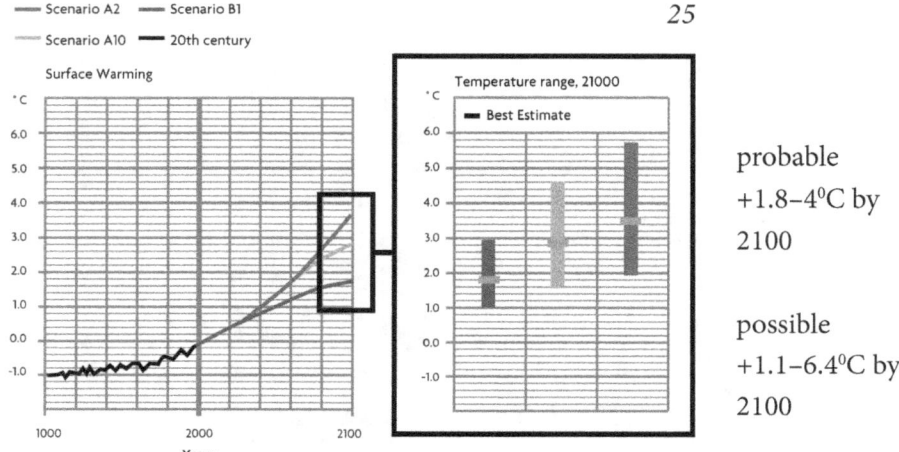

probable +1.8–4^{0}C by 2100

possible +1.1–6.4^{0}C by 2100

What reductions have been made in greenhouse gases?

METHANE
1984–2005

CH4 (ppb)   *26*

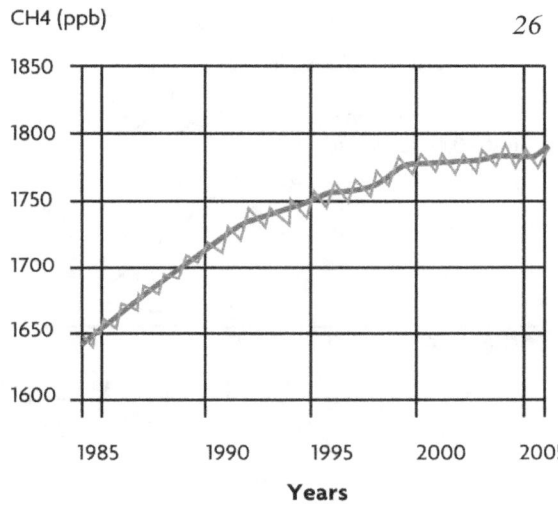

What has come to pass?

Expected 2.5% per
decade arctic ice loss

Predicted +0.15–3⁰C per
decade global temp rise

NITROUS OXIDE
1988–2005

N₂O (ppb)

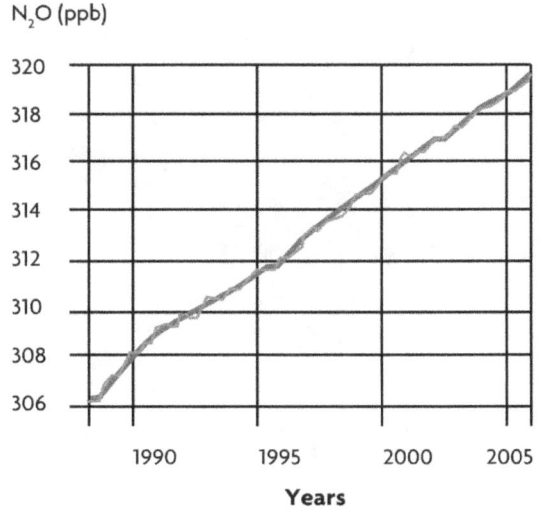

Observed arctic ice
reduced 7.8% per decade
over past 50 years

Observed +0.2⁰C per
decade global temp rise

1990   *27*

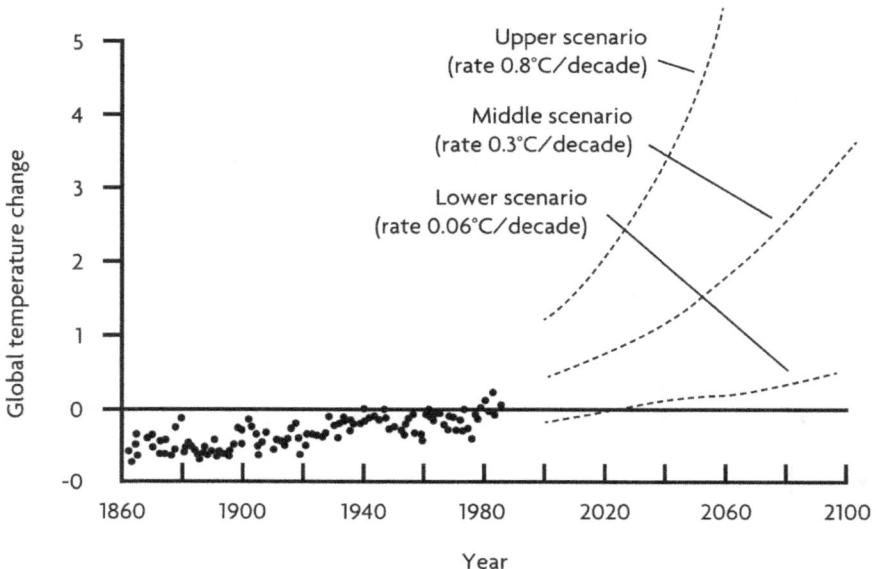

2007   *28*

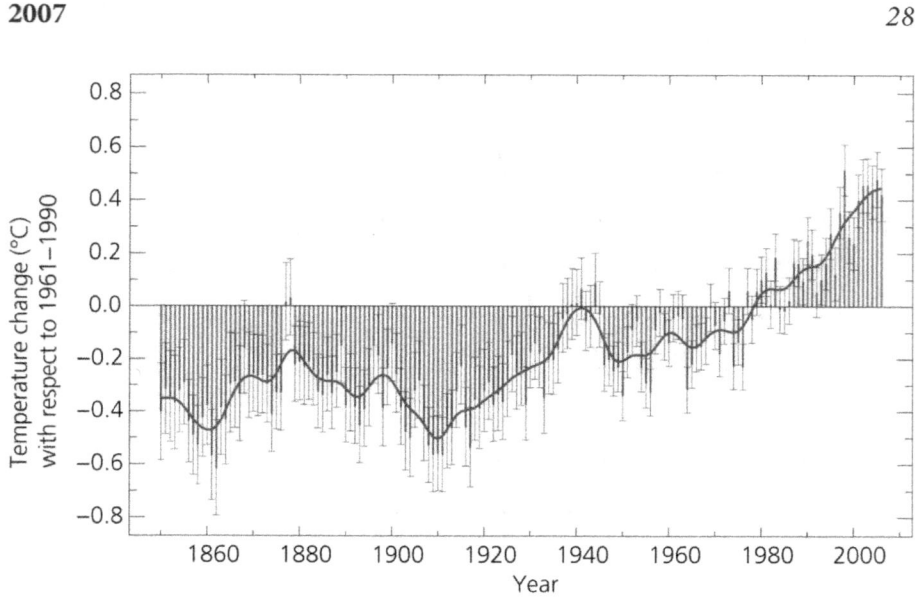

What action has been advised since 1990?

The original report did not even recommend freezing global CO_2 emissions at 1990 levels let alone cutting them, and despite the predicted global climatic disruption coming true in the past 17 years — hardly the blink of an eye in planetary terms — and the composition of a summary of possible mitigation strategies for this new report, the scientists *still make no specific recommendations to policy makers.*

Seventeen years of institutional inaction on the world's most important external issue — an issue easiest to remedy at its recognition and outset — whilst wars were planned and fought over fossil fuels.

The Joy of Ambrosia

"When you understand, reality depends on you. When you don't understand, you depend on reality."

– Bodhidharma

Failing all that, I have one last hope: that I might make people so depressed about the state of the planet that they stay in bed all day, thereby reducing their consumption of fossil fuels.

– George Monbiot[1]

My work aims to have the opposite effect to that of George Monbiot's. If you are depressed, by definition you will have no energy. If you have no energy, you will be unable to fulfil your direct responsibilities of life, work and family. Your chances of simply existing will begin to start registering in the negative. Your chances of having a positive effect on society, indeed the planet, at large, will be registering solely in the negative.

In contrast, I want my reader to be left feeling active and empowered for the challenges ahead, enthused at the possibilities for change and in their natural state of looking forward to the future with positive expectation.

With this position, goes another message:

"Use your energy"

And another *modus operandi*:

"Do more, Use more"

Life is an upward spiral of growth and all growth requires energy.

Life is an upward spiral of growth and all growth requires action.

This book recounts my own spiral of growth on a global journey. It catalogues my own personal growth and the philosophy expounded herein does not occupy a single position, rather charts a changing understanding that grew in the tale and in the telling.

It is a story about ignorance: mine own.

It begins at the time in my life when I was at my most ignorant, at my most isolated. The reason for this particularly acute spell of ignorance and isolation, paradoxically perhaps, was that I had recently achieved knowledge of a different order, and I had taken from this experience a very deep existential conceit. I knew things other people did not know, I understood things other people did not understand, I had abilities other people did not have; and I *knew* it.[2]

Revelation

As such, it is a story about revelation. A story about revelation and its consequences, and it catalogues the beginnings of the lifetime journey of dispelling my own ignorance.

"the joy of lucidity is at first like poison..."

"when they find, they will be disturbed..."

Revelation in the first instance, far from ideas of bliss and angels, is one of the most disturbing things that can ever happen to you. Irrefutable subjective proof that there is more to life than the immediate five sensory experience is about the biggest shock you can get. It changes *everything*.

That there is actually something to the religious ideals you were told about as a kid, those ideals that we half-carry, half-question, and continually lie to ourselves about, *is* shocking. That there is a way to contact and experience the profound, but that this understanding has been withheld from us, corrupted from us, is even more shocking.

Belief Systems

Scientific and religious belief systems form the foundations of our minds. I arrived in Sydney shortly after my scientific and religious belief systems had been annihilated by that extraordinary experience; consequently my mind was unstable: I no longer had any foundation to stand on.

Rather than affirming our childhood tales and succouring us with the nectar of external salvation, nothing in fact destroys belief systems like a genuine spiritual experience. Revelation is a fundamental challenge to existing religious beliefs, and an outright destruction of scientific ones. This is especially disturbing as we do not consider scientific beliefs to be just that; beliefs. We consider them as the rock of truth and reality. Reality, however, is not objective and it is not rational.

Australia

It is a story that took place in Australia.
Australians may be offended at some of my attitudes towards their culture and nation. I was most certainly not enjoying myself when I first arrived. But Australia is a place I came to love; a land of exceptional beauty, diversity, wonderful natural advantages, vibrant cosmopolitan culture and a high quality of life. My frustration with elements of life there was born of concern — I did not want those 'sustainable population' estimates to come to pass.

Indeed, my enthusiasm for the topics discussed in this book is due to the benefits they can potentially bring to Australian and all other societies. With its combination of environmental fragility, abundant sunlight, high technology and a highly educated populace as well as trade links to Asia, Australia is ideally placed to be a world leader in all of the fields of sustainable solutions; from land regeneration and sustainable agriculture to ecodesign, complementary currencies and renewable energy. Moreover, the contrast of the world's most internal culture living alongside the world's most external culture offers the opportunity of unparalleled cross-cultural growth and learning should there be reconciliation and dialogue. I wish every Australian a bright future.

Real Life

It is a story about my own spiritual journey, but also the story of a completely mundane journey.

It was just real life. For large parts of the trip I was merely getting by and making what I could of the conditions I found myself in. On occasion my own personal weaknesses and insecurities are on show. At times I was struggling, at others I found it easy and straightforward, at others I had an unbelievably great time. Names have been changed to protect the innocent, and some details have been omitted due to personal choice, or may be inaccurate due to the vagaries of memory — but overall it is pretty much exactly as it happened to me. Real. And this is important because the mundane and the spiritual are not two.

It is a story about sustainability, and a story about mortality.

I am still a relative beginner when it comes to training my mind, cultivating my *qi* or contemplating "Sustainability", but I am sure what I have learned will be of value to you.

Critical Responses Anticipated

There are some main criticisms which I can anticipate. Professional Buddhists will take issue with my equivalence of the Hindu *atman* with the Buddhist *tathagatagarbha*. I would respond that these words are just that; words; words which are a metaphor for *gnosis*, and I am aware that they do not represent some kind of object that can be grasped. Some will also take issue with my equivalence of emptiness with awareness. I would respond that the answers are found in direct experience, and not in ideas. Many understand emptiness of independent existence as a concept, few can transmit the experience the concept was developed to describe directly to others — or have experienced it for themselves.

"Doctrines are only for pointing to the mind"
– Bodhidharma

Professional philosophers will no doubt spend little time in pulling apart my use of terms such as "free will". I would respond that everything depends on assumptions.

For those who accuse me of spreading Mahayana Buddhism by stealth, I would respond that I am not a Buddhist. I am a human being from my particular part of the world, with my particular experience of life. What else can a human being ever be? I have no religious position and my philosophy is one of Self-Realisation.

In my discussion of the great tragedies and suffering of the world, historic and contemporary, juxtaposed with the nature of the metaphysical reference material I have included I will be accused of the high immorality of blaming the victim. I would respond that it would be immoral of me to remove all personal power from either my reader, my critics, or from the countless statistical beings which populate discussions of this nature — from the reports of global institutions to that of their NGO opponents — by rendering them as not responsible for their own existence.

A related criticism would be the promotion of the rational cardinal sin of "magical thinking". In my experience you can think either in limited terms or unlimited terms; the choice and results are yours. Beliefs beget thoughts and emotions beget actions beget results.

Because of the nature of the books I have referenced in the metaphysical sphere, I am well aware that I am inviting heavy criticism and ridicule. It would have been completely possible for me to produce a book for the 'mainstream', with all spiritual or metaphysical content removed and all my argumentation shifted into purely rational terms.

But this would not have been the truth of my experience.

Furthermore, this occlusion and denial of subjective experience in the scientific and intellectual spheres is the great failure of modern westernised culture.

I will be accused of childishness and sentimentality in my discussion of the development of my own convictions. I would respond that childhood is an important part in the journey of life and that sentiment is important in the discovery of personal truth; the suppression of emotional and intuitional knowledge and the open discussion of such another consequence of the historic masculine and rational emphasis of our western roots.

Psychological Impact

This is a story about global problems.

By far the most difficult aspect of looking deeply into global problems is the psychological impact the studies bring. My own voyage of attempting to cure global warming was a journey through what I came to identify as two highly deleterious states of mind:

As I attempted to educate myself on the greater issues of the day, my concept of the world grew increasingly darker and darker and the possibilities for the future correspondingly bleaker and bleaker. The gap between what I expected to happen and what I wanted to happen grew.

However, the world itself was not necessarily growing darker nor the future bleaker; this dead future was essentially just a picture in my rational mind; a consequence of the particular abstract data I was assimilating.

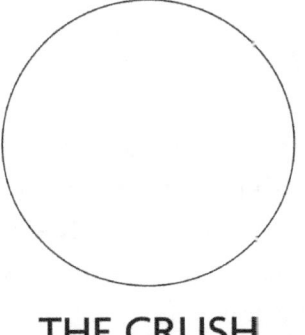

THE CRUSH

My subjective experience at that time was a result of my projecting the extrapolated worst consequences of this data before me into the future.

This is the rational approach to abstract problem solving. I have described it as "The Crush". The worst extreme of this state of mind is to live in a dead future, where all hope becomes false and all action futile.

Related to this was the attempt to reign in all of my behaviour in order to have no perceived negative impact on the world. An impossible task which simply generates frustration, which is psychologically hidden and internalised, then projected onto others if not consciously recognised.

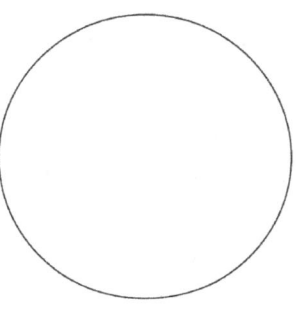

As noted before, I came to view all of my decisions as being increasingly moral in nature, **ECOPARALYSIS** with increasingly deleterious consequences for the long line of people on the other side of those choices. The degree of moral self-judgement and condemnation you feel in this state is mirrored by a corresponding condemnation of the world at large for its temptations and ongoing judgements on the actions of others.

This state of mind results from perceiving action as a negative, and conceiving of 'reward' as success in avoiding action. This is the essence of asceticism; an inversion of creative and existential actuality.

Personal Growth

The aim of this book is to dispel ignorance. The ignorance coincident with the primordial separation we experience between the world 'in here' and the world 'out there'. Dispelling ignorance through promoting awareness; greater awareness of self and of world, and of the importance of communicating our inner life:

Awareness (knowledge + conceptual understanding + life experience) + Communication

= Creativity + Empowerment

With the intended outcome of encouraging the personal empowerment and actualising the creativity of the reader. In this way it aims to reproduce the benefits of my own journey, and to encourage the reader to set out upon a journey of their own.

Note the breakdown of what I have defined as "Awareness". Before I travelled I was already educated to degree level and had experience in industry. I had a lot of what we would usually consider as 'knowledge and understanding' of the world at large. However, I discovered that in many ways my understandings were two-dimensional, requiring the missing element of actual life experience to make them real.

The other missing element was what I have termed "conceptual understanding":

As an example, we all would say we understand money. We use it every day and in many ways our lives depend on it. But after reading the work of Bernard Lietaer or Deirdre Kent, how many of us can say they have a conceptual understanding? How many of us can manipulate the very concept of money in our minds multi-dimensionally; to look at it from different angles, appreciate the assumptions therein, trace out the full range of effects and their dynamic? Conceptual understanding means understanding a concept systemically, or holistically.

Crucially, without both (i) adequate knowledge and (ii) life experience; (iii) conceptual understanding is unlikely to be achieved.

Single Issues?

This book aims to solve global problems.

To cure by identifying the root, rather than palliatively treating the symptoms.

As an antidote to the overriding public focus on single issues, I have included a series of overlapping reference works on the major issues of the day, works which have that rare ability to provide a complete conceptual shift in understanding of their topic in question. Interconnected topics require interconnected understandings. Furthermore, by reading through all of them, you will have the opportunity to gain a superior total perspective than any of the individual authors had at the time of their writing. Paradigmatic assumptions are often invisible and the whole is greater than the sum of its parts.

This book, together with its extended bibliography, functions as a reading list for those who wish to come to terms with the challenges of the new millennium.

Sunyata

If it has introduced you to the Mahayana Buddhist concept of **Sunyata** then it has achieved its main goal. Once you experientially understand this concept then all things will become transparent to you. **Sunyata** brings the intuitive into the rational. It allows us to understand the interconnections. When we understand the interconnections, then we can act in accordance with life.

We can ascertain the ecology of ourselves and our world.

We can keep our house in order.

According to the multidimensional logic of Sunyata

According to the multidimensional logic of **Sunyata**, all questions can be answered as:

- ☐ Yes
- ☐ No
- ☐ Neither Yes nor No
- ☐ Both Yes and No

i.e. Question: *is atman (self) the same as anatman (no self)?*

- ☐ Yes — both descriptors of spiritual nature of our being
- ☐ No — logical opposites; you either have a self or you don't

---

☐ Neither Yes nor No — just words conveying ideas rather than direct experience

☐ Both Yes and No — different philosophical ideas {no}, pointing at the same experience (which cannot be described in words) {yes}

We can find different people holding each of the above points of view — just as we can hold each and all of these points of view ourselves at different times — and they are all the correct point of view :reality: to the person holding them, at the time of their holding them. With *Sunyata* we can comprehend all of these points of view, and hold all of these differing perspectives in mind simultaneously.

With *Sunyata* we have the ability to understand ourselves more fully and in context, to understand others more fully and in context; all the while recognising the relativity, individuality, uniqueness and difference of perspective of every being in the universe.

Wu-Xing

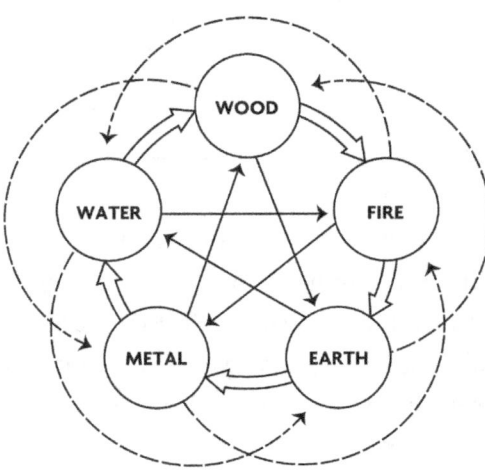

Activating and controlling relationships in a self-regulating whole system

With a whole system approach comes a tracing out of the inter-relationships, as inspired by the ancient Chinese concept of the five phases; *wu-xing*. My critique of the "Green approach" is based on the degree that it occludes systemic understanding, and neglects the importance of inter-relationships in sustaining a whole. A holistic conceptual understanding on the other hand, appreciates the inter-relationships.

Systems

With *Sunyata*, goes a famed refrain:

"Form is Emptiness, Emptiness is Form"

A refrain which demonstrates the importance of that which is considered Green Sin: Energy.

A down-to-earth example provides illumination:

Employment is Energy, Energy is Employment.

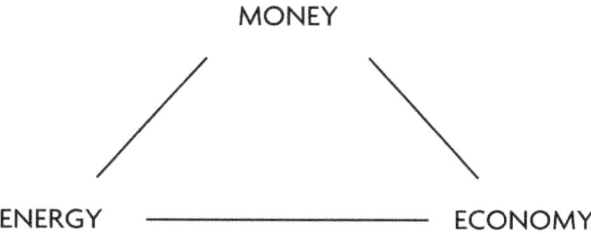

MONEY

ENERGY   ECONOMY

To reduce the net energy of a society is to reduce its employment, is to decrease its possibility of feeding and clothing its members. Attempting to bring about a sustainable society by pushing directly against the necessities of daily existence is bound for failure. A campaign against energy is a campaign against the ability to go to work in the morning. This is energy as sin, rather than energy as the flow of life itself.

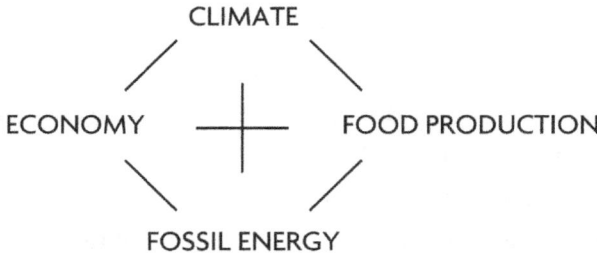

In contrast, highlighting the importance of energy through promoting an understanding of the inter-relationship with daily life is much more likely to catalyse positive change in the form of an energy transition. Likewise, highlighting the importance of ecology and climate through promoting an understanding of their inter-relationship with daily life is much more likely to catalyse positive change in the form of a monetary transition.

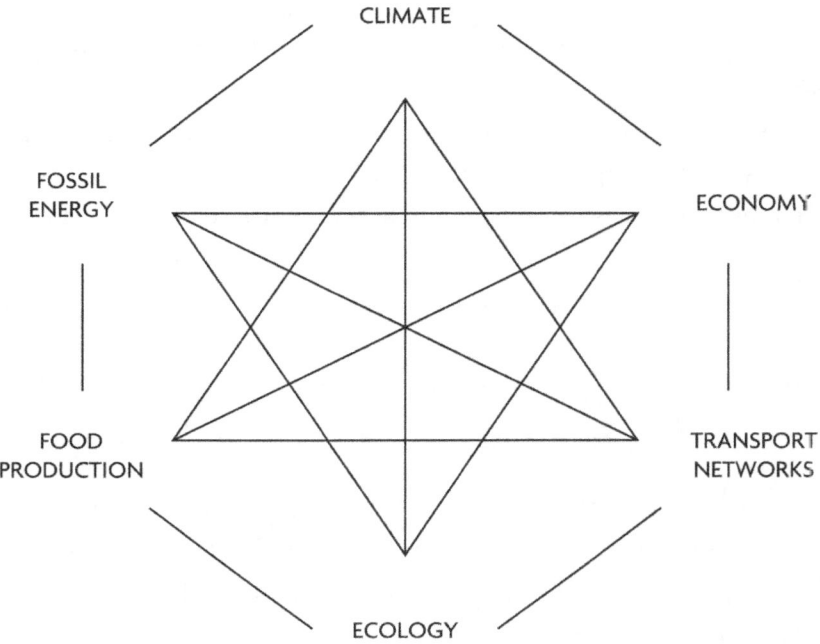

Ecoliteracy

In essence, this book aims to promote a greater literacy; an eco-literacy. In that sense, it is a return to philosophy. I find it quite staggering to reflect on the fact that, in our modern society and in our modern age, it is possible to be educated to degree level and yet not be exposed to the works of the greatest minds of all the ages. Why is this the case? Perhaps because the works of these minds teach us *how to* think, and not *what to* think.

Which leads us to the biggest question: Why no change already?

It seeks to answer the biggest question I encountered when travelling across the globe.

When I set out on my world tour, I already had a lifelong interest in environmental problems and environmental solutions behind me. Looking out at potential unrealised, the most pressing question brought to mind was — **with all these problems already known, and all these solutions already available, why was so little being done to solve them?**

I think that the answer comes down to a combination of these main factors:

(1) Don't know   (2) Don't believe

(3) Not practical   (4) Against the interests of those currently in power

Don't know
Many people in the world do not have the benefit of the literacy, technological communication, knowledge and education we in the developed world have. As a consequence there may be a lack of understanding of both environmental problems and environmental solutions.[3]

Don't believe
People on encountering this knowledge may choose to believe or not believe in it, based on any number of subjective factors including (i) how relative to their

day to day lives is it? (ii) from this new understanding, what changes would be necessary within their lives? (iii) what benefits are to be gained by action as opposed to non-action? and so on.

Not practical

Human beings are intelligent and basically practical by nature and invariably choose the most effective and efficient means of life available. The main reason why many renewable and alternative energy systems are not currently viable is that they are outcompeted by the storability, portability, availability, power-to-weight ratio; *ergo* efficiency (economy) of fossil fuels. This is the technological hurdle.

Against the interests of those currently in power

The forth issue is one of political power and control. Those who control fossil fuels will be threatened by competing energy sources, those who control money supply will be threatened by new forms of currency, those who own vast material assets will be threatened by land reform, those who monopolise political power will be threatened by institutional reform, and so on.

That these groups take measures to protect their own interests is well documented.

"There is nothing more difficult to carry out, nor more doubtful of success, nor more dangerous to handle, than to initiate a new order of things"

The, what could be more vernacularly termed "Turkey's-voting-for-Christmas principle", has been identified as the biggest barrier to social change by every social analyst before, during and after Machiavelli.

But we cannot blame our governments, and we cannot blame our leaders. We are the government, and our leaders are expressions of ourselves.

If we wish to spread a new paradigm of eco-literacy throughout global society, we can work on (1) and (2) using the communication means already available, (3) is technology dependent and requires investment and (4) is the most difficult to change, requiring a tipping point within the mass understanding of society as a whole to be reached before manifest change is possible.

The way to reach this tipping point is through the giving of knowledge (1) and the changing of perceptions (2).

Historically social change has been a bloody process, but the "Global Energy Transition + Path of Sustainability" will be an evolution rather than a revolution, based on enlightened self interest, communication and free will and not on legalism and the coercion of external authority.

Detachment

This book contains a touch of *Zen*. It is a rediscovery of the discoveries of the ancients, and it is inspired by their wisdom. It aims to introduce their way; the way of training the mind.

> *"Detachment is the essence of the way"*
> *–Bodhidharma*

This book looks at ominous and terrifying large scale issues. It looks at the large scale events which threaten and seem to define our lives. But to merely react in accordance to incoming data is to be in danger of living our life as a puppet of circumstance.

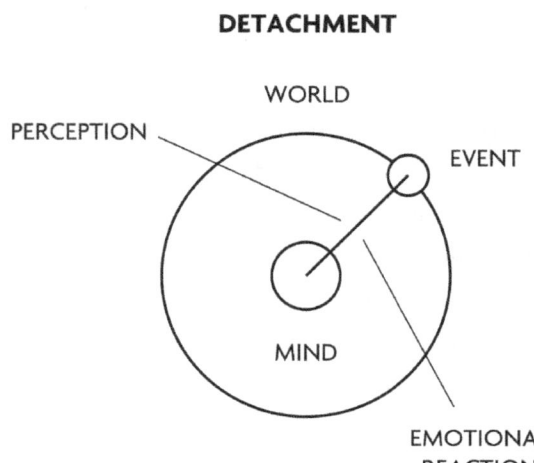

We cannot disconnect our rationality from our emotions. But we can view our self with detachment.

With detachment, we can be self aware. With detachment, we can respond to circumstances, rather than react to circumstances. We can live life proactively.

What We Think We Become

"The thought manifests as the word;
The word manifests as the deed;
The deed develops into habit;
And habit hardens into character.
So watch the thought and its ways with care"

If you don't want to accept any nebulous metaphysical *modus operandi* behind thought, then simply take my arguments figuratively. Action follows thought. Change the thought and you change the action. This is accepting and accessing our power as living, self aware sentient beings.

Worldview

Every self is a world. Every individual mind is a worldview and the paradigmatic assumptions for each worldview are invisible to that individual.

Perception is reality. One man's prophet is another man's false prophet. One person's true religion is another person's false religion.

Perception is not only reality; perception is a filter. Our perceptions filter the data we take in from the world at large.

A focus on the problems, disaster and tragedy of the world leads to a downward spiral of perception. Rationality is always associated with emotion, and so without detachment and self-awareness abstract problems can generate great fear. This emotion depresses the mindscape, stresses the body and darkens the thoughts. Limiting beliefs form around this contracted mindscape, which in turn further filter data and further bias perception. Even more frightful data is gathered, which increases time spent in negative emotional states, in turn increasing negative ideation. And so on, and on, towards nihilism.

This is the great danger of relying only on the rational mind.

Emotional Psychology

The Crush

The emotion underlying "The Crush" is Fear.

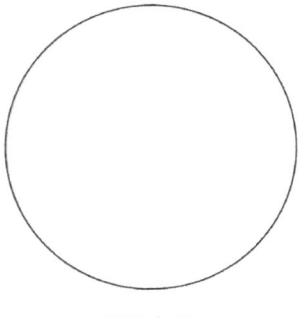

FEAR

"The Crush" is permanent fear of the future.

Fear associates with destructive and escapist thoughts.

The object of fear must be either pushed away or escaped from.

At its extreme 'The Other' must be destroyed.

Symptoms of living in permanent fear are trying to isolate yourself from the world, or always running away from the problems you encounter in life.

Ecoparalysis

The emotion underlying "Ecoparalysis" is Guilt.

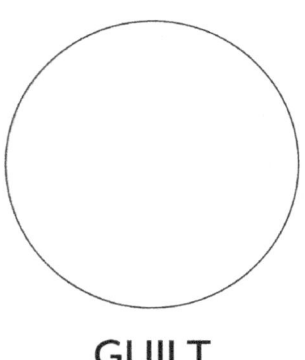

GUILT

"Ecoparalysis" is to feel guilty about something you are either dependent on or can do nothing about:

Inescapable guilt.

Guilt associates with self-destructive thoughts. The self must be punished.

At its extreme, the self must be destroyed.

The symptom of living in permanent guilt is a complete loss of sense of humour. It leads to a life lived in permanent self-justification.

If these abstract states of mind seem familiar to you, it is because they are the basic tools of religious and cultural conditioning.

Despite fossil fuel use being such a foundational part of my travel experience, I only began to seriously reflect on my journey's CO_2 impacts after my arrival in Fiji, prior to this ongoing motion was simply a necessary and therefore invisible assumption of my decision to travel. The ideas of Fiji and the UK obviously being poles apart in my mind, a guilt was brought up which resulted in me brooding deeply during my time there; a state of mind which only affected myself, and anyone who came into contact with me, negatively and which did not change the daily necessities of life in any way at all.

Throughout my time in Fiji, however, there was concrete evidence of the **morality** of my everyday actions; due to the coup taking place, packaged travel to Fiji had crashed with only adventurous or foolhardy backpackers being there in significant numbers. The air of despondency and depression which permeated the country during my time there was a direct consequence of this cutting off of business. With less work and less money people were finding it increasingly difficult to simply exist on a daily basis. Then I realised; why on earth was I feeling guilty? It will always be possible to go to Fiji. The "moral or ethical" removal of fossil fuels, whether by voluntary non-travellers or systemic mechanism, to the transport network connecting the Islands to the rest of the world could only result in the "moral or ethical" death of much of the population of Fiji.

These are the two occult egocentric hurdles of contemplating sustainability and the planet. The other is the more usual kind of egotistical superiority; when it comes to knowledge of the world at large, being more informed than the average person has the potential to indefinitely inflate your own sense of self importance.

Nonconceptual Awareness

Critics of *Zen*, with its emphasis on non-though, claim it to be anti-intellectual, when in fact it is a process of *enhancing* the intellect and expanding the mind.

The term "Awareness", as I have employed it in this book, does not correspond exactly to the awareness spoke of in Buddhist texts. These most often refer to the non-conceptual awareness of an empty mind.

("Awareness" in the context of this discussion is used in an effort to describe the expansion of consciousness that comes from the combination of training the mind and of engaging fully with life.)

Paradoxically, the non-conceptual awareness trained in meditation increases the capabilities of the rational mind for conceptual thought. It enhances the intellect.

New Consciousness

Zen is spontaneity, and spontaneity comes from within. Meditation awakens the intuitions, and only when the rational mind is supported by, and couched in, the intuitions does the waking ego not live in a constant baseline fear of death with its associated existential paranoia and unconscious survival strategies.

When rationality and intuitions work together and not in opposition there a great expansion in human consciousness; one enhances the other and both are improved. You are improved and the world is improved because of this. A new **State of Being** is realised; more precisely, a **State of Becoming**.

Conceptual Shift

I prefer the term 'conceptual shift' for the sorts of changing understandings that can come from reading a book. Indeed "Paradigm shift" is perhaps the most over-used phrase in modern progressive literature. But I can do better than just talk about paradigm shifting: I can give you the keys to the door. If you follow the metaphysical references I have given, books like *Chi Kung for Health and Vitality* and *The Nature of Personal Reality*, you will have the opportunity to gain, for yourself, direct experience of a whole new paradigm of human existence. The choice of walking through that door, and the experiential path of daily meditation, is up to you.

A word on genuine internal arts; there is a massive degree of difference between someone practicing genuine *Raja Yoga*, which is training of energy and mind, and the physical stretching exercises which are now found in gyms everywhere. Likewise, there is a massive degree of difference between the genuine *Shaolin Wahnam Qigong* I practice, which is training of energy and mind, and the gentle calisthenics found in parks, and syndicated on DVDs, everywhere. The difference between a high level internal art and a low level non-internal art is demonstrated in their **efficacy**. Test any claims made with your own direct experience.

Of course there is always the potential for fakery and exploitation in traditions which are not culturally mainstream and in skills which are not visible; when one Asian mystic publishes a best seller a thousand and one new 'teachers' spring up in the tourist traps of the east. The difference between a genuine master of the arts and a faker is **morality**; perfect ethics comes from perfect intention.

Right Attention

Zen is right attention.

When faced by problems, right attention is a focus on solutions. This is quite a contrast to the overwhelming focus of our culture, which could be described as omnipresent habitual wrong attention. Our entire intellectual culture, as any glance at the newspapers and newswires will tell you, is problem oriented. The problem orientation, of course, is rooted in the fact that we are all mortals.

Solution Orientation

The importance of a solution orientation can be seen in the literature I have referenced. James Lovelock's *Revenge of GAIA* and *The Final Energy Crisis* edited by Andrew McKillop are problem-oriented looks at global warming and fossil fuel limits respectively, and are two of the most depressing books ever written by human beings. Consequently they are unlikely to catalyse the public response to these problems their creators were presumably hoping for.

In contrast, Deirdre Kent's *Healthy Money, Healthy Planet*, and Jeremy Leggett's *Half Gone* cover many of the same topics, but are solution oriented. These books are uplifting and inspiring looks at the challenges presented, and are much more likely to leave the reader filled with enthusiasm for the road ahead.

Crucially, many of the solutions referenced in the latter two books are <u>not incorporated in the discussions of the first two</u>.

This is due to the problem orientation creating a **perception bias**; assumptions filter the data that is incorporated into a worldview. These perceptions cannot be separated from emotion, emotion which further limits perception in a downward spiral of negative attention. This leads towards a rigid, limited and dark worldview.

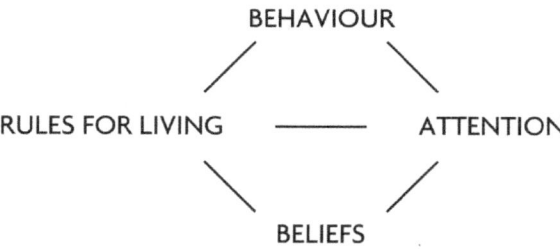

On the Future

The future is not James Lovelock's couple of "breeding pairs" huddling around camp fires in the arctic, nor is it Andrew McKillop's entropic nightmare of certain mass "die off". The future isn't in their mind, my mind or your mind, because the future does not exist and never exists.

Only the present exists.

Everyone is capable of dealing with their personal present.

It was only when writing about my journey that I came to appreciate the wisdom of the ancients; to take best care of the personal present <u>is</u> to take best care of the future, personal and global.

As the ancient Chinese knew, the cultivation of global harmony is implicit in the cultivation of personal harmony. By keeping our own house in order we play our best part in sustaining the global house.

"Let everyone sweep in front of his own door, and the whole world will be clean."
– Goethe

Feasibility

Regarding feasibility, every specialist weighing in on the "Sustainability" debate, from the Peak Oil school, to the pockets of solar, nuclear, wind enthusiasts, and so on, have their own arguments and predicted outcomes, sometimes with wildly different statistical data and assumptions to back up their predictions, sometimes extrapolating different effects from exactly the same data.

Feasibility exists in one place: the mind of the beholder. No one is capable of assimilating all of the available data nor of predicting the future — and that is why I have not attempted to back up all of my statements with statistics and instead went for the broad strokes of potential.

Graphs, Charts and Statistics

The world is far too complex to be contained by graphs, charts and statistics. It is impossible to define a subatomic particle by these means, yet we find it appropriate to use them to describe *A World*. High level graphs, charts and statistics, those used as evidence in discussions of this kind, are composed of figures which are single static representations; single static representations of what are actually floating multidimensional variables, multitudes of floating multidimensional variables which are each in turn dependent on multitudes of contributory parameters; variables which can change radically as even a single contributory factor changes. The greater the scale encompassed by two dimensional graphs, charts and statistics, the further they are certain to be from reality; the higher the level of the statistics used by global agencies, the less representative they are to life on the ground. Approaching feasibility by considering graphs, charts and statistics as rock bed reality is bound to fail.

They are useful tools at best, so long as it is understood they can never capture living reality, and they should never supplant real life experience in gaining an understanding of real world conditions.

Moreover, there is the danger of emotional disconnection from what statistics represent. People are neither percentages, nor are they determined effects.

Feasibility Conclusion

On the contrary, I have approached feasibility in terms of that which is most concrete and immediate; beginning from what is possible with one's own house, here and now, and working from there outwards.

But what of other people's houses? Is our own house not constantly endangered by the hopeless disorder of others'? This fear is most often to be found in the population debate.

On Population

The population debate — the modern arena where some of the world's most distinguished scientists, academics and weighty intellectuals can reference their fellow human beings in terms of casual dehumanisation, without any danger of public reproach.

To worry about world population is a road that leads to nihilism. To intellectually take on the responsibility for the lives of billions of people is to be mentally crushed under the weight of the inconceivable numbers. Population responsibility begins and ends with your own gametes.

Everyone is capable of this.

In contrast to the rational fear of uncontrollable exponential overpopulation leading to certain mass annihilation, all of the evidence of the developed world would suggest that population rates will drop towards zero as living standards improve. Indeed, the demographic trends of the west suggest the rising of the opposite problem of an aging and under-populated society.

But what of Malthus in Africa? To promote such events such as the recent genocide in Rwanda as a primarily demographic and environmentally determined result, and therefore to implicitly suggest that all societies are essentially 'ticking time bombs', is to gravely misrepresent the realities of life on earth.

Malthusians in Africa?

Rather than viewing situations such as the Rwandan civil war as the inevitable result of Malthus' dilemma, we should instead ask; to what degree are these conditions brought about through the work of Malthusians?

Structural adjustment prevents human beings from mobilising energy and resources.

To attempt to understand the 1994 massacres in Rwanda, we must look at the events in their full historic colonial and neo-colonial political and economic contexts. Whilst there were undoubted demographic and environmental inputs into the situation, we should also question the extent to which these were effects as well as causes.

The German colonists in the late 19ᵗʰ century used the native royalty to put into place a system of military outposts to control the region. Their Belgian successors in turn used the royalty as colonial enforcers and procurers of forced labour, and a social division along ethnic lines was largely fabricated and accentuated as part of a 'divide and conquer' strategy of domination. A polarised society and a climate of fear and distrust developed. The Belgians also put into place a land ownership system of individual plot cultivation geared towards a cash crop export economy. This system largely remained in place post independence, but despite it social progress was made in the 1970s and early 80's prior to the onset of the international debt crisis. When the main cash crop, coffee, suffered a market collapse spurred on by first world producers in the late 80's, the stage was set for the IMF-WB's 'bitter economic medicine' in 1990. Simultaneous with the beginning of the civil war, standard structural adjustment procedures including currency devaluation resulted in inflation and the collapse of real earnings, expanding external debt, the bankruptcy of state and public enterprise and the disintegration of health and education services. A further devaluation in 1992 resulted in further price increases of essentials, whilst farm gate prices of coffee were frozen at 1987 levels despite international highs. These factors had the consequence of virtual collapse of the economic value of the major crop, and therefore the inability of local producers to buy food; a situation further exacerbated by the rising cost of farm inputs and the destabilisation of internal markets by foreign food aid.

Concurrent to the late 80's coffee crisis, IMF-WB loans to neighbouring Uganda, despite being policed to the last dollar in domestic expenditure, were ballooning into the U.S.-British trained and equipped Ugandan military, an arm of which was rebranded as 'the Rwanda Patriotic Army' (RPA). The stage was set for an 'army of liberation' to invade, with IMF-WB aid money fuelling the militaries on both sides of the civil war.

The mass mediated accounts of the genocide managed to largely ignore the colonial roots of the conflict and the actions of the Bretton-Woods institutions in creating both the economic crisis and military funding necessary for such a tragedy to unfold, as well as to completely obscure the intentional geopolitical machinations that resulted in the transformation of a Francophone country to an Anglo-American dominated one; Washington and its British allies established a neo-colonial foothold in central Africa[4], with the collateral result of there being less central Africans around to use up "scarce resources" before they have the opportunity to reach first world markets.

We also cannot remove from this picture all of the decision making of all of the individual people involved in all sides of the situation, up to and including RPA General Kagame's decision to shoot down President Habyarimana's jet. He did not have to do this.

Famine in the modern age

An interesting discovery when looking into the causality behind modern famine is that there is usually more than enough food in the affected region to feed everyone, but that this food is being exported to foreign markets.[5] This was the case in the 19th century potato famine in Ireland, of which my own ancestors had to flee. Ireland at that time was ruled by a foreign nation whose elite caste were happy to use Thomas Malthus' social theories as ideological justification and camouflage for decisions that put profit over people's lives[6], and this trend continues in elite circles to the present day.

Remarkably, global food availability is in fact notable for its abundance,[7] and local level food insufficiency is often a result of global oversupply; the dumping of deregulated grain surpluses — "food aid" — results in local level food cultivation becoming uneconomic.[8]

High birth rates occur in the world's most impoverished areas. High birth rates mostly occur where children are perceived of as a kind of mini-labourer in the struggle for existence. What these regions need is increased opportunity, education and employment; development.

Moreover, the ideas discussed in this book, such as negative interest currencies, new energy systems, organic agriculture[9] and ecodesign hold the potential for a new form of development that harmonises economy with its ecological roots.

On Poverty

There is more depth to that term than can be captured in an accounts sheet. Many of us, myself included, have experienced periods of good earnings and material assets and periods of little to no earnings and material assets and know very well that our intrinsic worth as a being did not change along the journey. Many people today are exceptionally rich, and I am grateful to them and wish them well — the personal material wealth of an individual is mirrored by the useful work that they do for society as a whole. My critique of the world's top heavy wealth is based on the degree of violence, exploitation and usury involved in its genesis, along with the flawed nature of positive interest money creation, and not on the fact that relative differentials in material wealth, which are perfectly right and natural, exist. Many people today live outside the cash economy and thus have *no money*, and so do not even register on usual poverty statistics — these people are not necessarily living in need of salvation, and making them dependent on outside sources creates new problems. There is no reason why the CEO of an international corporation should have the same means available as someone who lives a hunter-gatherer existence in the rainforest un-contacted by civilization. Both are their own self-made individual expression of life and neither is better or worse than the other.

One of the happiest places I ever visited was the village of Baishi near Lijiang in the uplands of Yunnan province in China. Those people were most probably financially poorer than you, but the sense of harmony, peace and happiness in that village was real and tangible and not based on romanticism. With beautiful countryside, clean water, productive traditional agriculture and deep and strong personal relationships those people were rich. Solving the extreme poverty of famine is one of the reasons I wrote this book, but there is another extreme poverty which is closer to home in the heart of every city in the first world; a poverty so extreme that people's relationships are degrading and they are losing the very thing that defines life — Consciousness.

Downward Spiral

A Cancer of the Planet?

Travelling across the great land surfaces of the earth fairly changes your perspective on population. Seven days on trains across Eurasia and three days on buses across Australia were notable for the almost complete lack of people. And after time spent in the wilderness the great metropolises of the earth are seen as what they are; oases of life and mind. London, Moscow, Tokyo,

Beijing, Bangkok, Singapore, Melbourne, Los Angeles; these are amongst the best places on earth to live, study, work or visit. The U.S. eastern seaboard, the north west of Europe, the Gangetic plain, the Yangtze basin, Tokyo Bay …the odds are that both you and everyone you know and love lives in such an "overpopulated" part of the world. There is a simple reason for this. The "overpopulated" places are the best places. Not many depopulation enthusiasts can be found living alone, in isolated shacks, in Siberia, though there is certainly plenty of space for them to do so.

Human beings are not a virus, disease or cancer of the planet any more than James Lovelock and his nearest and dearest are a virus, disease or cancer of the planet. Human nature is the nature of the earth and of life. We are the planet in expression. Our daily life does not exist in opposition to nature, but depends on nature.

Balancing human needs with environmental health will always be amongst the most challenging problems facing societies. It will always be a case of managing multiple divergent, diverse, competing and at times diametrically opposed interests. But it should not be considered impossible. If it was impossible there simply would not be 7+ billion of us here today.

Indeed, our current environmental problems are consequences of success and not of failure. Human beings are the consummate problem solvers, illustrated quite spectacularly by our modern numbers. That is more than seven billion minds to apply to the challenges of the new millennium and the responsibility, albeit unevenly, rests across broad shoulders.

Peace

Our societies constantly discuss problems, constantly discuss dangers. But as anyone who has travelled around the world will know, one of the most extraordinary things about life on earth is the complete lack of problems. Peace and stability is the experience of life. Almost everywhere in the world, day to day existence functions seamlessly. The peace and ease you are no doubt feeling right now, nestled happily at the centre of the universe, with life flowing by in the street outside, goes on everywhere. Moreover, it is an active peace and a dynamic stability that is never boring, ever changing.

When the peace of the world at large is described, people automatically call to mind the ongoing experience of war. But it has to be borne in mind that these situations are aberrations. It takes an awful lot of ignorance, and an awful lot of immorality to cause a war. In the example of the recent war in Iraq, it took the most senior and invested with authority members of western society *lying* — telling untruths — to their constituency regarding the presence of weapons of mass destruction. Not only this, it required the support, the ignorance and the apathy of the majority of the rest of society for the situation to come to pass.

When the peace of the world at large is described, people also call to mind the danger of criminals; thieves and murderers. How many times have you been attacked by a stranger in your life? Not many I am willing to bet. The reality of strangers is the reality of you and I. Is this really something to be afraid of?

The abuse of power by the few can only take place through the acquiescence of the many. The evil of war is caused by ignorance. Ignorance of the interdependent nature of reality.

The solution to ignorance is awareness. Awareness of the interdependent nature of reality.

Awareness (knowledge + conceptual understanding + life experience) + Communication

= Creativity + Empowerment

It will take an active, aware, informed and empowered citizenry to bring about genuine democracy on a world scale.

Global Government

Indeed, the march towards global government in a technological age holds the potential to become the most absolute and dangerous of tyrannies.

We don't need global government, but we do need global organisation.

The world is simply too interconnected to go backwards. We already have global organisation. The financial agents of global organisation; the IMF, the BIS, the WTO, the World Bank, must be reformed rather than dissolved.

These global organisations and our other supranational structures must be made:

- Democratic
- Accountable
- Transparent

Member nations' participation should be voluntarily and not mandatory. These agencies must be separated from oligarchic interests. They must be separated from corporate media agendas. They must be separated from military intrigues, espionage and enforcement. They must genuinely come to represent global interests. Their constituency should be the whole world.

A new paradigm of genuine global cooperation must be born.

Authority

This new paradigm is a change to Authority. Individuals should know themselves as the authority in their lives. Our sovereignty should be the sovereignty of the Self. Outer authority, the authority we invest in our organisations and their agencies, should be brought back to its original meaning:

Invested responsibility as service.

Where once authority was the voluntarily elected and invested leadership of the community, it was corrupted over the ages so that the symbols of authority became more important than service to be rendered, and the control enabled by the authority more important than the lives of others.[10]

Holarchy

It is also a change to hierarchy. In a hierarchy each level of organisation is superior to those below; the parts are subsidiary to the whole. A holon on the other hand is where each part of an organisation is known as both a part and a whole, and where communication flows both 'up' and 'down'. Holon's are self-complete networks embedded one inside the other in an overall holar-chy. While they may exist at differing levels of organisation, they are all equal in importance. The metaphor was developed by Arthur Koestler to describe the human body; where individual elements — cells — meet their own needs through multi-levelled cooperation yet also are aware of and contribute to the needs of the overall whole. But of course such understanding of natural organisation has been held by other societies for a very long time.[11]

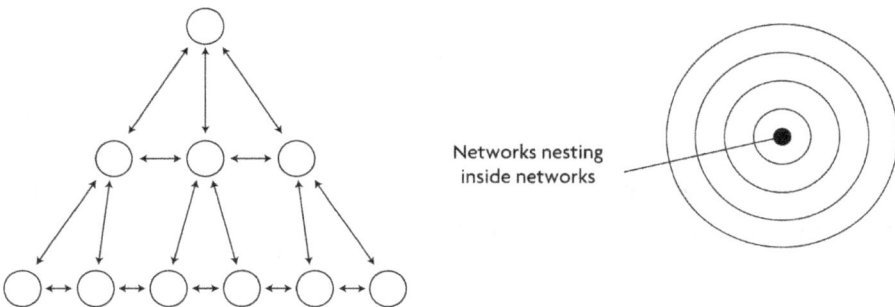

Networks nesting inside networks

It is this pattern of organisation that enables nature to allow many parts to act together as one greater unit. The origin of the term democracy is *demo*; people, *kratos*; power. When people are recognised as both wholes and parts, all equal and important, when the levels of our organising together are recognised as both wholes and parts, then all of the people can be more fully empowered, and in their individual empowerment enable the more full empowerment of one another.

Sustainability

"Sustainability". Another over-used buzzword. Sustainability is in the mind of the beholder. I have used the term Sustainability in this book to mean, as Viktor Schauberger did; *copying and learning from nature*.[12] Sustainability under this definition is about building cycles. It is about learning from the feedbacks and self-organising systems and processes of nature and applying these principles consciously to benefit all aspects of our society — organisation, finance, trade and industry as well as agriculture, aquaculture and environmental management.

The goal of a sustainable society should be what Bernard Lietaer calls *sustainable abundance*; to create a society that satisfies its needs without diminishing the prospects of future generations, whilst simultaneously providing freedom of choice and creativity to as many people as possible.[13]

Seeds of Change

With the glacial pace of institutional change and preponderance of vested interest in both the open and closed corridors of power, it is clear that such changes will have to come about through people working from the bottom up rather than the top down.

Thankfully, there are a great many of such projects at all levels already bringing sustainable solutions into being;

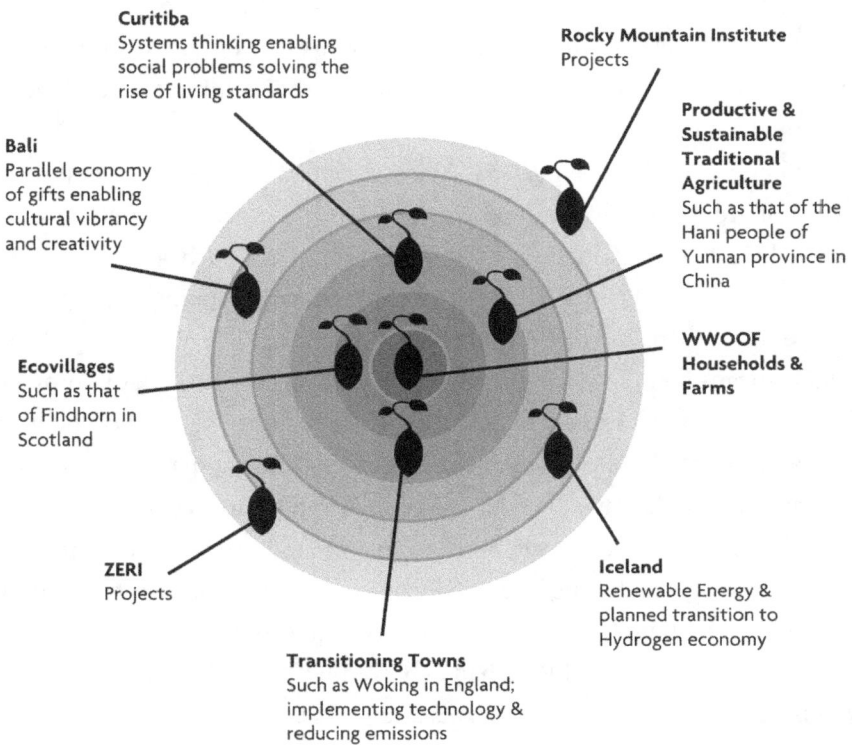

Curitiba
Systems thinking enabling
social problems solving the
rise of living standards

Rocky Mountain Institute
Projects

**Productive &
Sustainable
Traditional
Agriculture**
Such as that of the
Hani people of
Yunnan province in
China

Bali
Parallel economy
of gifts enabling
cultural vibrancy
and creativity

**WWOOF
Households &
Farms**

Ecovillages
Such as that
of Findhorn in
Scotland

ZERI
Projects

Iceland
Renewable Energy &
planned transition to
Hydrogen economy

Transitioning Towns
Such as Woking in England;
implementing technology &
reducing emissions

Key: 1. Personal 2. Community 3. Town & County 4. City 5. Nation 6. International

With the WWOOF movement we have many householders and farms utilising renewable micro-power, ecodesigned buildings and sustainable agriculture; the Zero Emissions Research and Initiatives (ZERI)[14] projects are pioneering cyclical resource flows and solving problems at all levels of society based on natural design principles; the world's traditional agriculture, such as that of the Hani people of Yunnan province[15], provides multitudes of examples of productive and sustainable solutions based on nature; the Ecovillage movement, such as *Findhorn*[16] in Scotland have created first world communities founded on sustainable principles; the Brazilian town of Curitiba brought the third world to the first in one generation by applying systems thinking to governance and issuing an alternative currency[17]; the Rocky Mountain Institute of Colorado is pioneering a broad spectrum approach to energy transition including hydrogen fuel cells and next generation automobiles[18]; Bali's traditional gift economy runs parallel to the standard economy and is responsible for the cultural creativity and community to be found on the Island[19]; the country of Iceland currently generates 80% of its energy from renewable sources already and is aiming to close the gap by switching to a Hydrogen economy by 2050[20].

Wherever goals have been set, there has been both learning and results. Towns all around the world have independently set goals to become carbon neutral by 2050 or sooner, as have whole countries such as New Zealand, Norway, Bhutan and Costa Rica. By 1998 the town of Woking in the U.K. had cut its emissions to 29% of 1990 levels and generated 43% of its energy by sustainable means including combined heat and power (CHP), solar photovoltaic (PV) and fuel cells. The town of Vaxjo in Sweden aims for carbon neutrality by 2030 and now 90% of home heating comes from biomass. Samso Island in Denmark has already achieved carbon neutrality through wind power and biomass. Cities like Vancouver, Stockholm and Freiburg are employing a variety of approaches including energy efficiency, solar power, investment in new technology, integrated mass transit and recycling to achieve their goals. By 2006 the country of Costa Rica had achieved 94% of its electrical generation from hydroelectric power, wind farms and geothermal energy.

These successful pilot projects contain many of the necessary ingredients for a transitioned society, and exemplify what is already possible. Moreover, they and others like them are already seeding further change all around the world. But even more such pilot projects are needed to test and develop the most ambitious ideas of sustainability, such as legal tender negative interest currencies put in place with contemporaneous changes to land ownership, legal systems and local governance.

Empowerment

The sun provides enough energy to power world society. I am not suggesting that the way ahead is necessarily to coat the Sahara in solar thermal installations. This is just an illustration of how much energy is available from renewable sources. But if you will take a minute to join me in a dream we can see the potentials; silicon and light from the sun is harvested in, thus enriching, those areas which are traditionally most poor; deserts. It is used to split seawater into hydrogen. Hydrogen is used as cellular fuel around the globe. Each of these cells are connected into a network magnifying the energy available to the world. Every geographic region makes use of its own renewable energy and combination of micro and macro potentials. Building design becomes ecological, manifesting the diversity and uniqueness of the regions of the earth. With currency reformation the energy, creativity and intelligence previously dormant in people through artificial scarcity is unlocked. Wealth flows through society, rather than stagnating in debt bubbles and hoardes, at all levels; locally, nationally and globally. Empowered, people can solve their problems from the bottom up. Long termism and cooperation become the norm for economic actors. Regeneration of the environment and sustainable use of natural resources becomes the sensible choice for nations. A more active and enabled society can better communicate. With more and better communication comes better organisation. New emergent local, national and global organisations arise to represent this new co-operative grassroots order. Solarised, the world will become much more stable, much more democratic and much more fair. And the future will look after itself.

Of course, the future cannot be predicted, but the potential of solving all of our major environmental and social problems is there — and with currency reformation it almost becomes a matter of course.

To take an active role in creating such change requires us to empower ourselves.

You cannot be empowered with a negative attitude to power.

You cannot be empowered with a negative attitude to energy.

The Good News

Global Warming, growing global debt, depleting fossil fuels, the potential of ecological and social collapse, the potential of more war and terror. This book certainly contains a lot of "bad news". But it also contains what has historically been called "The Good News". News so good it can never be expressed on a page.

> *"The Tao that can be named is not the real Tao"*
> **– Lao Tzu**

> *"The way is wordless, words are illusions"*
> **– Bodhidharma**

It also contains solutions. Solutions which exist on the other side of our problems.

> *"If there is a problem you can't solve, then don't worry about it.*
>
> *If there is a problem you can solve, then don't worry about it."*
> **– Shantideva**

Solutions with the potential to enhance life for all.

A Change to God

The new paradigm is a change to God.

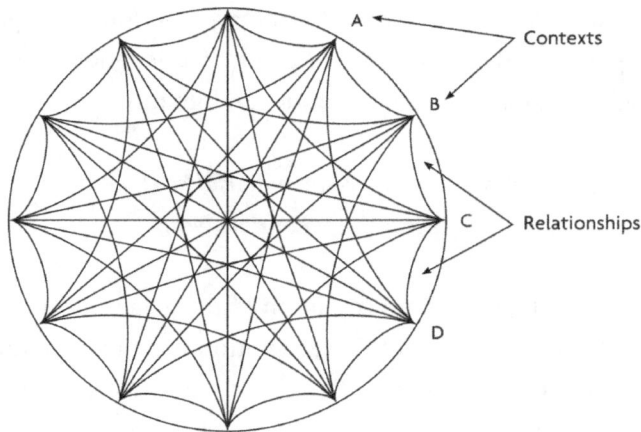

A worldview where the Divine is understood not as some kind of external interventionist entity, but as an immanent and transcendent ultimate reality; where the only method of communion is direct personal experience — *Gnosis*.

Duality and Non-Duality

The ancients challenge us to embark upon a path of self-transformation, from a worldview of separation to one of unity. Unfortunately, due to our cross-cultural heritage I have had to use the terms self and non-self to describe this, but duality consciousness and non-duality consciousness are better descriptors.

Non-duality does not mean elimination or sacrifice of the self. In fact, it can only be realised through expansion of the self; expansion of our consciousness to include the perspective of others, as equals. It is a natural result of the spiral of growth and expansion of the self that is following our intuition and fully utilising and developing our whole range of human capabilities. Contraction of the self on the other hand can only lead to a state of frustration, sickness and total obedience to externals.

Who benefits from this?

Creators

The new paradigm begins with us changing ourselves. *Gnosis* reveals us to be integral participatory components of a Divine Cosmos. As such, we must come to know ourselves as Creators in Creation — no longer as permanent sinners awaiting non-existent future salvation or chaotic random beings competing meaninglessly for survival in a chaotic random universe.

"Heart thinks, events materialise"
– Ho Fatt Nam

With this understanding comes a recognition of the generative power of our thoughts and emotions, and with this recognition an understanding of the necessity of greater conscious awareness of, and responsibility for, not only thoughts and emotions, but their origins; the contents of our mind.

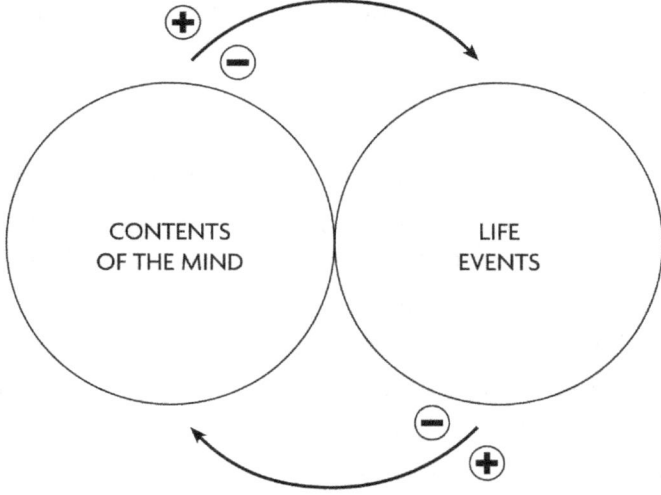

Relationships are mirrors of the self

This personal conceptual revolution of the foundation will initiate the changes we need to make in the outer world:

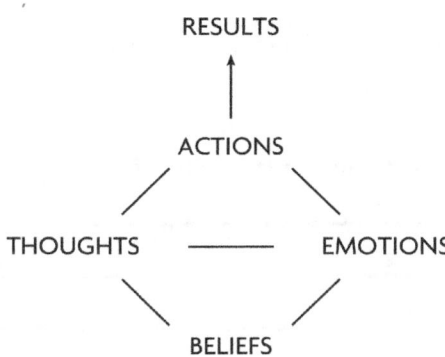

Personal Consciousness Revolution -> Social Consciousness Evolution -> Global Monetary Transition ->

1° 2° 3°

Global Energy Transition + Path of Sustainability

4° 4°

If we do not know the contents of our mind we must become aware of them, and if we do not like the contents of our mind then we must change them; and when we change them the world around us changes.[21] Our personal world mirrors our inner world. The mass world is a result of the combination of all of our inner worlds.

"We are what we think. All that we are arises with our thoughts. With our thoughts we make the world."

– Siddhartha Gautama

Wealth Gap

Indeed, rather than becoming shocked at learning of the widening gap between rich and poor; both within societies and between societies; we should instead be shocked if we see anything different happening — as it is a mathematical consequence of our money creation system. A money creation system of our own creation. A system based on beliefs in separation, superiority, competition, centralisation and conquest.

Net Interest Transfers

Systematic wealth transfer from the bottom 80% to the top 20% of society, Germany, 1982.

Corporate Globalisation

Status Quo:
Not conspiracy, consequence of our belief systems

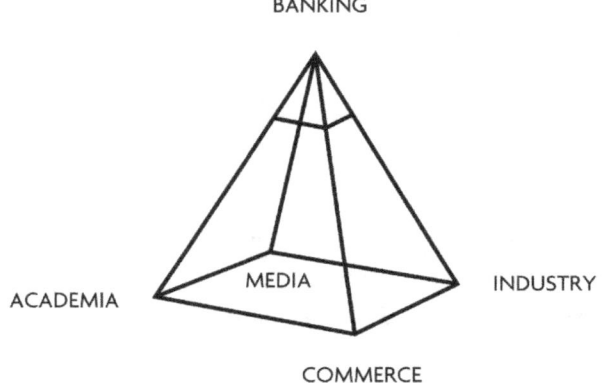

As for the neo-feudal global racketeering and loan shark arrangement known as 'Corporate Globalisation'; if the penny hasn't yet dropped here it is: the 20th century saw interest bearing fiat money go global and thus enabled the wealth transfer within nations to function between nations. Post WWII trade reform and the fall of Communism in turn allowed the world's now increasingly top heavy wealth to asymmetrically penetrate, through the vehicle of corporate shareholding, the majority of the resource base of the earth and the infrastructure of civilisation; the shareholders in central banks, the shareholders in commercial banks and the shareholders in the blue chip corporations that dominate global trade are found to be all of the same people.

Moreover, the revolving door between the think tanks funded by these global banking interests and the executives of government, the executives of supranational agencies, the executives of the intelligence services, the executives of multinational corporations and the executives of the military allows an unprecedented degree of cooperation at the elite levels of society and co-opts public executive towards the pursuit of their strategic objectives.

And that's one of the reasons why we live in a world with a handful of billionaires and billions with comparative handfuls.

Resistance

"Change comes from within"

The point of highlighting this detail is to raise awareness of the structure of global society. Such hyper-enrichment of elites can only happen when the people are asleep. But rather than continuing to focus on such asymmetries, or worrying about the power of such vested interests, the solution orientation requires a different approach:

No amount of complaining, blaming or worrying about global warming, the threat of terrorism or of engineered war, peak oil or any other major global issue will improve these situations. It will not change larger circumstances.

The solution orientation is to change focus once a problem has been identified and understood.

"Whatever is negative, whatever is self-centred, what feeds malicious thoughts or stirs up the mind, those states of mind draw one downwards; turn your attention away."

A continuing focus on the problems of the world will generate great fear and guilt. These emotions will simply weigh you down, bias your perception and distort your sense of the greater realities of life.

No amount of abstract fear will change the world. If you are feeling guilty about using fossil fuels or positive interest money, then likewise no amount of abstract guilt will change the world.

But it will surely affect you personally in the negative.

It will negatively affect your ability to simply exist. It will negatively affect your ability to fulfil your direct responsibilities of life, work, family and community. It will negatively affect your ability to influence the world at large.

If you take a stand against business as usual — as I did when I arrived in Australia — you will be crushed.

This is because you will be essentially taking a stand against life.

To complete a sustainability journey we must come full cycle; mountains must again become mountains and rivers again rivers. Those invisible unquestioned assumptions must go back to being invisible assumptions once more. But because they have now been questioned, we have increased ability to comprehend, and to act.

The rational approach is a problem orientation. This problem orientation is a resistant attitude to life.

On the contrary, as those wise Daoist ancients would say, you have to go with the flow...

Vs Acceptance

"Right attention follows from right effort. It means keeping the mind where it should be. The wise train the mind to give complete attention to one thing at a time, here and now."

Right attention begins with Acceptance. Acceptance of your self as you are.

Acceptance of your bad qualities as well as your good qualities is acceptance of the whole of you; acceptance of your mistakes and failures as well as triumphs and successes.

Acceptance is understanding that both polarities are necessary for the functioning of the whole.

Acceptance is understanding that you are acceptable as you are.

Zen is the here and now. To keep the mind in the here and now is to cultivate the optimum state of being. We must accept the here and now in order to keep our mind there.

To keep the mind in the here and now is to optimise our ability to live. It is to optimise our ability to respond to the direct responsibilities of life, work, family and community.

Right attention begins with Acceptance. Acceptance of the world as it is.

Acceptance of the bad qualities and good qualities, failures and successes of society is acceptance of the whole.

Acceptance is understanding that both polarities are necessary for the functioning of the all.

Acceptance is understanding that the world is acceptable as it is.

Acceptance is understanding that other people's lives are not problems to be solved.

ZEN

"The wise can direct their thoughts, subtle and elusive, wherever they choose..."

Meditation begins with the cultivation of a one-pointed mind, and aims for a mind at rest at zero.

A trained mind can expand, like **Sunyata**, to encapsulate abstract problems.

A trained mind can be focused, like a laser, and brought to bear on tasks in the here and now.

A trained mind improves our ability to conceptualise and to cognicise. To imagine and to visualise.

With a trained mind we can contemplate large scale issues without attachment. With a trained mind we can put our full attention on the here and now, and on what is important: the people and the relationships in our life. When we contemplate sustainability, we contemplate sustainability. When we work, we work. When we spend time with our family, we spend time with our family. When we play and have fun, we play and have fun.

"Whatever is positive, what benefits others, what conduces to kindness or peace of mind, those states of mind lead to progress; give them full attention."

Right attention is choosing to focus on what is positive.

What is positive in self, what is positive in others and what is positive in the world at large.

- Take a moment to focus on what is positive in the world around you. Write down all the good you see, feel and hear.
- Take a moment to focus on what is positive in your friends and family. Write down all the qualities you love, and all the fine and special experiences you have shared.
- Take a moment to focus on what is positive in yourself. Write down all your talents and skills, successes and achievements, good memories and experiences.

Breathtaking is it not?

"A trained mind brings health and happiness"

If you are resisting this exercise, then you lack love for something as important as the sun, the moon and the stars.

Love for Self

"You yourself, as much as anyone in the universe deserve your love and affection."
– Siddhartha Gautama

To love another, you must first love yourself.

"Love your neighbour as you love yourself"
– Jesus

Love exists to be expressed. How much love do you feel for the world at large? How much potential for expressing your love for the world at large do you feel?

Forgiveness

"to understand everything is to forgive everything"

If this expression is blocked then you may need to forgive. This is something only you can choose to do. It may be forgiveness of another. Or it may be forgiveness of yourself.

Appreciation

"Whatever is positive, what benefits others, what conduces to kindness or peace
of mind, those states of mind lead to progress; give them full attention."

Right Attention is Appreciation.
Appreciation of the here and now.
Appreciation of the self and of the world.
Appreciation is Gratitude. Stop to think about how much you have to
appreciate in:

• Your self
• Your family
• Your friends
• Your community
• Your country
• Your World

How much thanks have you given for all of these? How much thanks do you
have to give?

Reflection on learning

Appreciating the whole is appreciating your own personal learning journey,
and appreciating humanities learning journey;
    Take a trip to a museum and reflect on the lives past people led. Without
all of their work and all of their life things would not be as they are just now;
not only would we never have been born, but we would never have been able
to get clean drinking water whenever we wanted it, food whenever we wanted
it, warm water for bathing, heat through the winter, luxurious shelter, stable
democratic society, diverse choice of employment, ability to create our own
businesses, virtually unlimited choice of friends, virtually unlimited choice of
relationship partner, information technology, a world's worth of knowledge,
a world selection of products and services, a world's worth of creativity, art,

sport, entertainment, expression and fun. Many of us have never experienced exposure to the elements, hunger or malnutrition, war or devastation, dictatorial government, or restrictions on freedom of thought and expression. We have a lot to be thankful for.

Ten generations ago, rather than being a relatively young man with most of his life and professional achievements ahead, I would probably already be dead.

The time to live is here. The time to act is now.

A New Age

So why not seize the day? Think of things that have never been thought of before. Do things that have never been done before. Build things that have never been built before. Create things that have never been created before. Embrace the freedom and the potential and the power of being. The gift of life. Here and now. In this world.

We look back in awe at the individual human examples of the great sea change of consciousness that took place in the axial age three thousand years ago. Heraclitus, Socrates, Zoroaster, Mahavira, Siddhartha Gautama, Confucius, Lao Tzu. They have become legend. Their genius shaped our world.

At the turn of a new millennium, we are now experiencing a new axial age; an age where the axis of the globe itself is the axis of a single interconnected worldwide humanity.

The advantages of our age over those past peoples are beyond staggering, and these advantages are multiplied seven plus billion fold. An age where potentially billions of individuals, not just a shining few, can aspire to reach the same heights as our great predecessors.

We can go beyond, and we can uplift consciousness to a degree undreamed of by our ancestors.

We are currently living in the best of times. Our own lives are now brighter than many throughout history could ever have imagined. This is the best time as a human being to be alive and our decisions today will determine the future of the earth for all people, and all creatures.

So let's focus on what is good in life, with gratitude, and on the solutions to our challenges.

Regeneration

"The secret of health for mind and body is not to mourn for the past, worry about the future, or anticipate problems, but live in the present moment wisely and earnestly"

Nature, when we give her time and space to rest and recover, regenerates automatically. People impact on nature. But people also have the potential to work with the processes of life, to make the land and seas richer and more full of life. We have the potential to enrich the earth; to give as well as to take. Human beings themselves have incredible regenerative abilities not currently widely known or understood.[22] The future is potentially bright; far brighter than a dualistic mind can ever imagine.

Which leads us to the real question: if divine knowledge and experience is possible {gnosis}, why is this not generally known, understood or pursued?

I think that the answer comes down to a combination of these main factors:

(1) Don't know   (2) Don't believe

(3) Not practical   (4) Against the interests of those currently in power

Don't know
The Divine Within is not an accepted tenant of the major western religious traditions nor, perhaps more importantly, contemporary scientific thought. As such the experience is generally not sought, except after occurrences of personal crises or disaster.

Don't believe
People on encountering this knowledge may choose to believe or not believe in it, based on any number of subjective factors including (i) how relative to their day to day lives is it? (ii) from this new understanding, what changes would be

necessary within their lives? (iii) what benefits are to be gained by pursuing self-awareness as opposed to continuing on in ignorance? and so on.

Not practical
Meditation requires time alone, an aim of self-awareness and an attitude of self observance. In contrast the pace of modern life is rapid and exhausting, and its lifestyle of constant addictive stimulus is oppositional to time spent in contemplation.

Against the interests of those currently in power
The forth issue is one of political power and control. Those who provide the operating systems for the human mind (religion) will be threatened by any alternative spiritual movements that could change the *status quo*. Those who claim to represent objective truth (annihilationist science) will be similarly threatened.

If we wish to spread a new paradigm of *gnosis* throughout global society, we can work on (1) and (2) using the communication means already available, (3) is personal choice dependent and requires the investment of personal time and energy and (4) is the most difficult to change, requiring a tipping point within the mass understanding of society as a whole to be reached before manifest change can occur. The way to reach this tipping point is through the giving of knowledge (1) and the changing of perceptions (2).

Saving the World

"Save the World". The Planet does not need to be saved. Even asteroid impacts produce long term creativity rather than long term damage. Even worst case scenarios of human effects are not more damaging than such perfectly natural events. The Planet, and everything living on it, from plants to people, functions through self regulation. Blades of grass, trees and insects do not need you to look after them. They were here before you were here and they will be here after you have gone. Other people do not need to be "Saved". The people of the third world, for instance, are as capable of living their lives and organising their societies as you and I.

> *"No one else can save you.*
> *No one can and no one may"*

The world doesn't need saviors and it doesn't need to be saved. If you are having problems in your own personal life then only you can save yourself. If you aren't having problems in your own life and want to do something for others around; then do something positive:

> *"use more, do more"*

No one can "Save the World". But everyone can change the world. By changing themselves. If you want to effect major change on a Global Scale, then *Think Big*: Be Ambitious.

To play a part in reforming society on a global scale is the most challenging task that any individual can ever attempt, unlike the Green approach of replacing your lightbulbs, which is amongst the easiest.

The Secret

> *"The whole secret of existence is to have no fear; never be afraid of what will*
> *happen to you, depend on nothing. Only in this are you freed."*

Vast and terrifying though the problems facing us may seem to be, there is in fact no great evil facing the world. There is only ignorance.

Our own.

The solution to ignorance is awareness.

> *"Overcome anger by non-anger, overcome evil with good.*
> *Overcome the miser by giving, overcome the liar by truth."*
> **– Siddhartha Gautama**

This choice is in our own hands.

Love for All

"May all beings be happy."

Look into your heart and you will find love.

Love for yourself and love for the World. Every other heart is the same.

If they can't express this, then their joy has been veiled by ignorance and by suffering — not evil.

Initiating a New Order

The solution to ignorance and suffering is spiritual growth. To grow spiritually is to initiate a new order of personal being.

There is nothing more difficult to carry out, nor more doubtful of success, nor more dangerous to handle, than to initiate a new order of things.

Our ego seeks stability, safety, similitude and order, whereas spiritual growth is embracing and flowing with creativity, spontaneity and change.

For the reformer has enemies all who profit by the old order, and only lukewarm defenders in all those who would profit by the new order.

Our ego protects its own integrity by reacting against data which challenges it's most cherished and implicit assumptions about existence, and its current mode of living.

This lukewarmness arises partly from fear of their adversaries who have the law in their favor; and partly from the incredulity of mankind, who do not truly believe in anything new until they have had actual experience of it.

If you don't know about it, then it doesn't exist.

If you don't believe in it, then it doesn't exist.

Our waking ego requires proof that there is more to life than survival, and it is up to us to search for it; internally by training our mind and exploring our dreams, externally by embracing the challenges of life and turning those dreams into realities.

"The incredulity of mankind..."

There are big surprises in store for mainstream society.

Existing scientific belief systems are, putting it as mildly as I possibly can; incomplete.

"There are stranger things under Heaven and Earth, Horace, than are dreamt of in your philosophy"

– Shakespeare

Shall turn out to be the most prophetic words ever written in the English language.

Practical Problems

In conclusion, environmental problems are not moral or ethical problems. They are practical problems. Practical social problem solved by practical social means.

If we look at any of the fields where progress needs to be made in response to the challenges ahead, fields such as; sustainable land management, sustainable agriculture, sustainable aquaculture, land regeneration, sea regeneration, habitat protection, renewable and alternative energy, ecodesign, complementary currencies and grassroots democracy.

All of these fields require at the personal practical level:

"doing more, using more"

All of these fields require at the socio-political level:

"doing more, using more"

Do not "do less, use less". If there are social or environmental problems that concern you I recommend you take inspiration from the living example of those individuals who have been successful in the socio-political arena such as

Amory Lovins, who developed the Rocky Mountain Institute; Jeremy Leggett who founded the renewable energy company Solar Century; Jared Diamond's career as a world class academic, promoting the multi-disciplinary approach to the study of society and the environment; Jaime Lerner's career in politics as the mayor of Curitiba and beyond; or Deirdre Kent's lifetime of grassroots activism. This is:

"doing more, using more"

It takes energy to fulfil the direct responsibilities of life. And it takes energy to influence the world around. Influencing the world at the community, national and international level requires correspondingly increasing magnitudes of energy. This is as true in the 'Sustainability' sphere as in any other field of human endeavour. Everyone alive wants to leave the world a better place, so if you have a passion for life and a passion for positive change:

- *Use your Energy*

Think Big. Be Ambitious. Playing a part in change on a world scale dares you to dream and dares you to reach your highest potential in life.

"When you are inspired by some great purpose, some extraordinary project, all your thoughts break their bonds: Your mind transcends limitations, your consciousness expands in every direction, and you find yourself in a new, great and wonderful world. Dormant forces, faculties and talents become alive, and you discover yourself to be a greater person by far than you ever dreamed yourself to be"

– Patanjali

Global Transport Network

Expand your mind to picture the road leading from your house, spreading out around the world, leading eventually to everyone on the planet's house. Picture in your mind the industrial infrastructure necessary to power that network and the lives of everyone dependent upon it.

The problem is keeping that transport network running, not shutting it down. The problem is keeping that infrastructure functioning, not shutting it down. The problem, therefore, is continuing to use energy, not figuring out how not to use it. Moreover, in a time of global economic recession, transitioning global society is the greatest economic opportunity that could be realistically conceived!

There is no such thing as a Green object or a Green person.

Human beings, in common with all living things, are goal oriented, and negative goals can only lead to suicide.

Not using energy and resources is not sustainable.

If you do not use your energy and resources you will not be sustained. Neither will society.

We must come to understand that sustainability is not a quality of individuals; sustainability is a property of planets.

Doing Less

As for the reactive "Green" approach; combing the most important task that could ever be conceived; "Saving the Planet"; with the world's most disempowering message; "do less, use less"; is not going to work.

Detachment is required when reading the endless list of "do less, use less" Green solutions; solutions which often defeat the purpose of both intelligence, and indeed, existence. Detachment is required when observing its bizarre line in negative altruism. Detachment is definitely required when witnessing the constant media parade of this new global diet craze; the carbon aestheticism of multi-millionaire ecomoralists. Detachment is very definitely required when witnessing people in every country of the world scrimping and saving; "Going Green"; whilst the fruits of their labours are siphoned off to service the perpetual compound debt black hole of money which is not real and does not exist.

As is a sense of humour.

Social Engineering

While it may be occult to fanatics, the depressive effect of this "Green" message is more than appreciated by those sectors of society which do understand human psychology and which do plan long term; the social engineers of the technetronic era.

Despite the danger of systemic problems, and therefore systemic solutions, coming to light, the potential of ritually locking the masses into permanent fear of future and inescapable guilt was too good to miss, hence the global championing of the "Green" approach.

The original original sin corrupted a whole civilisation of human beings for millennia, and the Green equivalent has the potential to do the same to the whole world for millennia more. Environmentalism as a global religion — with its saved and sinners, judgement and punishment, and salvation or enlightenment through self-constriction — is a snug fit into existing religious belief systems.

When human beings do not feel worthy or capable of living their own lives they turn to an outside authority for salvation.

Central organisation is always more than happy to play this role.

"...it is a campaign not for abundance but for austerity.
It is a campaign not for more freedom but for less.

Strangest of all, it is a campaign not just against other people, but also against ourselves."

– George Monbiot[24]

It is easier to limit and control people when they cry out for limits and controls to be placed upon them. It is easier to perpetually drain resources and energy from people when they do not believe they are either worthy or capable of using them.

So unlike the overwhelming majority of Green literature out there I do not recommend any kind of "doing less, using less". Mass outbreaks of what amounts to a combination of obsessive compulsive disorder and agoraphobia are not going to save the planet. They will not change it in a positive manner. If you are not concurrently "earning less, spending less" then all this amounts to is saving money. And saving money is bad for both the economy and the environment.

All this amounts to is a life lived in psychological conflict. And if you are feeling guilty about your own existence, then you have automatically been rendered unconscious of the degree to which you are structurally politically and economically disadvantaged. Wake Up.

On Travel

This book recounts an experience of world travel. But it can never truly capture the reality. How do you transmit the experience of a world? Life on earth is an unfolding series of wonders and marvels; an unfolding series of wonders and marvels that simply defy my ability to capture in writing. It cannot be described. It can only be experienced. *Go experience.*

Try travelling far and wide without a particular destination in mind. Travel, as I discovered, is the most *Zen* of experiences; where you're not attached to the past and you don't worry about the future. You just appreciate the present moment.

Whatever your means you can decide today to do something different; to go somewhere you've never been, to do something you've never done, to meet someone you've never met. Take a step outside your front door and you're on a moving network that connects you with everyone alive, and everyone that's ever been alive. The opportunity is always there. The power is always yours. *Please travel.*

Please use energy. Please aim to live a full and complete life. Please aim to reach your highest potential.

To see the world is to know that we are living in paradise. Not in the past, not in the future. But in the now. It is only now.

> *"The returning road is very long and also short,*
> *The returning road from Shaolin is tortuous and also even.*
> *What are you thinking all along the way?*
> *The breeze so gentle, the sky so blue.*
> *A drop of water will not dry in a river.*
> *Creativity and success all start with your first step.*
> *Cleanse the dust from your heart; the whole body becomes light and happy,*
> *Only in the happiness of other people can you find your greatest happiness"*
> — **Yan Chen**

End Notes

Thanks

Thanks to Mother, Father, Patrick, Christopher, Sean, my three favourite ex-girlfriends, "the guys" I grew up with, and everyone I saw and met along the way of my 2005/6 journey; especially to the gracious families affiliated with the WWOOF movement who welcomed me into their homes and inspired with their *eco logical* ideals.

Thanks to Deborah Bird Rose for allowing my excerpt of her work *Dingo Makes Us Human*, and thanks to Roger T. Ames for allowing me to mercilessly plagiarise his understanding of the western and Chinese paradigms. Thanks to Clive Ponting, Richard King, Bertrand Russell, Niccolo Machiavelli, James Lovelock, Ross Gelbspan, Deirdre Kent, Greg Palast, John Perkins, Jared Diamond, Angelo M. Vernados, Jamal R. Nassar, Swami Parmhans Maheswaranda, Jeremy Leggett and the Dalai Lama for their disembodied thoughts and companionship on the road. Thanks to all of the other featured authors for allowing me to interface with their work and thoughts through the vehicle of their writing once I returned home.

Thanks to my teachers Jane, Rob, Rick, Esther, Jerry, Sanaya and Duane.

A very special thanks to my Sifu for his lifetime of work spreading health and happiness to people irrespective of race, culture or religion — and for transmitting the skills and experience that made this book possible.

Thanks to you dear reader; may your most heartfelt dreams come true ...in this life.

And thanks to ...you know who you are.

Notes:

Opening:

1. Source: Tom Rees http://www.brighton73.freeserve.co.uk/gw/paleo/paleoclimate.htm

Compiled from temperature data from Mann et al, 1999, CO_2 data from Law Dome ice core Paleoclimatology Branch of the National Climatic Data Center, and Mauna Loa Carbon Dioxide Analysis Center.

Arrival:

1. Lovelock Quote: http://www.independent.co.uk/voices/commentators/james-lovelock-the-earth-is-about-to-catch-a-morbid-fever-that-may-last-as-long-as-100000-years-523161.html

2. The Story So Far; Glasgow to Sydney via Russia, Japan, China, Tibet, Vietnam, Cambodia, Thailand, Laos, Malaysia & Singapore.

3. Cyclone Larry, see NASA http://earthobservatory.nasa.gov/NaturalHazards/view.php?id=16268

4. Are you a Human Being or a Light Switch? The dualistic nature of perception.

State of the World:

1. Source: Chalabi & Mellon, *Wake Up! Survive and Prosper in the Coming Economic Turmoil*, Capstone Publishing Ltd. (2005), Fig 2.1, pg 64; based on OECD data.

2. CREDIT: Association for the Study of Peak Oil and Gas (ASPO) http://www.peakoil.net/uhdsg/weo2004/TheUppsalaCode.html

3. CREDIT: November Coalition http://november.org/graphs/http://en.wikipedia.org/wiki/File:US_incarceration_timeline-clean.gif

4. Source: Food and Agriculture Organization of the United Nations (FAO) data, http://www.fao.org/fishery/topic/3016/en

5. Source: International Geosphere-Biosphere Programme (IGBP); Steffen *et al* (2004) http://www.igbp.net/

6. CREDIT: National Oceanic & Atmospheric Administration (NOAA), Earth System Research Laboratory data

http://www.esrl.noaa.gov/gmd/obop/mlo/

7. CREDIT: UNEP/Grid Arendal

http://www.grida.no/graphicslib/detail/world-population-development_29db

Corporate Globalisation:

1. Quigley, *Tragedy and Hope: A History of the World in Our Time*, New York: MacMillan (1966)

2. Bank for International Settlements (BIS) History,

http://www.bis.org/about/history.htm

3. BIS Ownership,

http://www.bis.org/about/charter-en.pdf

4. BIS Purpose,

http://www.bis.org/about/index.htm

5. Wood, *Global Banking: The Bank for International Settlements*, The August Review (2005)

6. BIS Protocols,

http://www.bis.org/about/protocol-en.pdf

http://www.bis.org/about/headquart-en.pdf

7. Peters, *The Bilderberg Group and the project of European unification*, Lobster magazine (1996), http://www.lobster-magazine.co.uk/issue32.php

8. For Blair failure to declare:

Hansard Commons Answers

http://www.publications.parliament.uk/cgi-bin/newhtml_hl?DB=semukp arl&STEMMER=en&WORDS=bilderberg%20blair&ALL=Bilderberg%20 Blair&ANY=&PHRASE=&CATEGORIES=&SIMPLE=&SPEAKE R=&COLOUR=red&STYLE=s&ANCHOR=muscat_highlighter_ first_match&URL=/pa/cm199798/cmselect/cmstnprv/180iii/sp0304. htm#muscat_highlighter_first_match,

For an exchange which is misleading, if not technically untruthful (Blair was a member of the Opposition during his 1993 attendance), see also:

Hansard Commons Answers

http://www.publications.parliament.uk/pa/cm199798/cmhansrd/vo980330/ text/80330w06.htm

9. Korten, *When Corporations Rule the World*, Kumarian Press (1996), pg 138

10. Wood, *The Global Elite — Who are they?*, The August Review (2005),

http://www.augustreview.com/issues/globalization/ the_global_elite%3a_who_are_they?_200511146/

11. http://www.augustreview.com/images/pics/tcm-2006.pdf

12. Palast, *The Best Democracy Money Can Buy*, Penguin (2003), pg 153

13. *Ibid.*, pg 154

14. McKillop *et al*, *The Final Energy Crisis*, Pluto Press (2005), pg 102

15. Palast, (2003), pg 154–155

16. Chussudovsky, *The Globalisation of Poverty and the New World Order*, Global Research, Centre for Research on Globalisation (2003), pg 47

17. Chussudovsky, (2003), pg 49

18. Palast, (2003), pg 155

19. *Ibid.*, pg 145–146

20. *Ibid.*, pg 156

21. *Ibid.*

22. United Nations Development Panel (UNDP), *Human Development Report 1996*, http://hdr.undp.org/en/reports/global/hdr1996/

23. UNDP Chief, James Gustave Speth, quoted upon release of 1996 Human Development Report. http://www.mtholyoke.edu/acad/intrel/incomgap.htm

24. Wolff, *Recent trends in wealth ownership 1983–1998*, (2000)

25. Source: data from Wolff (2000)

26. *Ibid.*

27. http://news.bbc.co.uk/1/hi/business/4619189.stm, http://www.globalpolicy.org/socecon/develop/debt/2005/01payingforrelief.pdf

28. http://www.unodc.org/unodc/en/data-and-analysis/WDR-2005.html

29. http://www.globalsecurity.org/military/world/spending.htm

30. Brzezinski, *The Grand Chessboard*, Basic Books (1997), pg 27

Escape from Kings Cross

1. An example of Ecodesigned Housing.

Return to Sydney

1. See Grossman, *On Killing: The Psychological Cost of Learning to Kill in War and Society*, Back Bay Books (1996)

2. Plutarch's description of the Spartans (see Russell, *History of Western Philosophy*, Unwin Hyman Ltd 1984, pg 120)

GAIA:

1. Lovelock (quoted in Capra, *Web of Life*, Harper Collins, {1996}, pg 203)

Sick Money:

1. Quoted from Frederick Morton, *The Rothschilds, A Family Portrait*, New York: Atheneum (1962)

2. Source: Lietaer, *The Future of Money*, Random House (2001)

3. *Ibid.*

4. Growth Curves

5. Dennis Meadows, Jorgen Randers, Donella Meadows, *Limits to Growth: The Thirty Year Update*, Chelsea Green (2004)

6. Kent, (2005)

7. Quoted from the documentary, *The Money Masters: How International Bankers Gained Control of America*, (1996) www.themoneymasters.com

Brisbane:

1. Extended quote from *The Bhagavad-Gita*, translated by Barbara Stoler Miller, Bantam Dell (2004)

Little Korea

1. Diamond, *Collapse: How Societies Choose to Fail or Succeed*, Penguin (2005), pg 398

2. McKillop et al, *The Final Energy Crisis*, Pluto Press (2005), see Chapter 20 *France and Australia After Oil* by Sheila Newman

3. http://www.abs.gov.au/AUSSTATS/abs@.nsf/allprimarymainfeatures/73DC AFE12953D30ACA257134000AD22B?opendocument

4. Diamond, (2005), pg 397

Ecological Interdependence:

1. Easter Island: Colonisation

2. Easter Island: High Point

3. Easter Island: Descent

4. Easter Island: Contact

5. Tikopia

6. Tokugawa Japan

7. Malthus' Dilemma

Yoga:

1. Karma Diagram

2. Union Diagram

Global Warming:

1. Legget et al, *Global Warming: The Greenpeace Report*, Oxford University Press (1990), pg 12

2. Source: UK Department of the Environment, *Global Climate Change*, May 1990

3. *Ibid.*, pg 13

4. *Ibid.*, pg 30

5. *Ibid.*, pgs 29–41 , pgs 44–67

6. Source: *Ibid.*, pg 20

7. Source: J. Jaeger, *Developing policies for Responding to Climatic Change: a summary of the discussion and recommendations of the workshops held in Villach, 28 September to 2 October 1987*

8. *Ibid.*, pg 13

9. Source: D.A. Lashof & D. A. Tirpak, *Policy options for stabilizing global climate*, U.S. Environmental Protection Agency, 1989

10. *Ibid.*, pg 99

11. *Ibid.*, pg 105

12. *Ibid.*, pg 480

Outback

1. See Kakadu National Park, *Park Notes* (NM 06/05), published by Australia Government, Department of the Environment and Heritage. See also:

http://learnline.cdu.edu.au/tourism/kakadu/values/pdf/aboriginalart.pdf

Mysticism de-mystified

1. See Wong, *The Complete Book of Shaolin*, Cosmos (2002), Chapter 26

2. See Wong, *The Complete Book of Zen*, Element (1998)

An Introduction to Buddhist Philosophy

1. *Samyutta Nikaya*; translation quoted from King, *Indian Philosophy: An Introduction to Hindu and Buddhist Thought*, Edinburgh University Press (1999)

2. *Ibid.*

3. Quoted from *The Zen Teaching of Bodhidharma* translated by Red Pine, North Point Press (1989)

Creating a New Paradigm: An Ecological Paradigm

1. Capra, *Web of Life*, Harper Collins (1996), pg 41

Validation:

1. Ames, *Sun Tzu The Art of Warfare*, Ballantine Books (1993), pg 113

2. Brzezinski , *The Grand Chessboard,* Basic Books (1997), pg 35

3. *Ibid.* Pg 36

4. *Ibid.* Pg 25

5. *Ibid.* Pg 211

6. R. Kagan, P. Wolfowitz, W. Kristol *et al, Rebuilding America's Defenses* (2000). http://www.newamericancentury.org/RebuildingAmericasDefenses.pdf, pg 63.

7. http://www.newamericancentury.org/iraqclintonletter.htm (1998).

8. Brzezinski *The Grand Chessboard*, pg 148. My addition in parenthesis for clarity.

9. *Ibid.*, Pg 35. My addition in parenthesis for clarity.

10. *Ibid.*, Pg 53–54

11. McKillop, Newman *et al, The Final Energy Crisis*, Pluto Press (2005). See Chapter 5 *The Caspian Chimera* by Colin J. Campbell.

Unlimited Energy:

1. Source: NASA http://solarsystem.nasa.gov/multimedia/display.cfm?IM_ID=2166

2. U.S. noon average, http://www.ucsusa.org/clean_energy/technology_and_impacts/energy_technologies/how-solar-energy-works.html

3. Geothermal gradient; http://en.wikipedia.org/wiki/Geothermal_gradient

4. I am aware of differing scientific theories of solar function and solar-earth relationship, but even speaking conventionally the existing potentials can be seen.

5. Earth's Magnetic field.

6. Source: http://www.theoildrum.com/node/2583

7. McKillop *et al, The Final Energy Crisis*, see Part I.

8. World Organism Diagram.

9. Again, I am aware of differing theories of hydrocarbon generation, but whatever the physical limits, an end to the burning of hydrocarbons (CO_2 emissions) and their wise stewardship as the material basis of modern society (cyclical use) would still be an aim of a path of sustainability.

10. McKillop *et al, The Final Energy Crisis*, see Part I. (pg 85)

11. Hawken, Lovins & Lovins, *Natural Capitalism*, Little Brown, New York 1999, pg 35–7

12. CREDIT: Aubreymeyer; http://en.wikipedia.org/wiki/File:Web_C&C.png

Paradox I

1. Monbiot, *Heat*, Penguin (2006), pg 172

Reconnection:

1. Leggett, *The Empty Tank: Oil, Gas, Hot Air, and the Coming Global Financial Catastrophe*, Random House (2005)

2. Lietaer *The Future of Money*, Random House (2001), pg 43, 51–2, 242–4

3. *Ibid.*, pg 27, 246–8

4. *Ibid.*, 249–59

5. Douthwaite, *The Ecology of Money*, Green Books (2000)

6. Source: Lietaer (2001), pg 248

7. *Ibid.*, pg 27, 246–8

8. Brzezinski , *The Grand Chessboard*, Basic Books (1997), pg 3

9. http://gata.org/node/5738

Epilogue:

1. Dawkins, Richard *The God Delusion* (2006), pg 117

2. *Ibid*, pg 112

3. Davies, Paul *The Goldilocks Enigma* (2006), pg 4

4. *Ibid*, pg xiii

5. See http://www.thirdworldtraveler.com/Global_Secrets_Lies/Global_Slave_Trade.html, also http://www.foreignpolicy.com/articles/2008/02/19/a_world_enslaved

6. http://www.todaysalternativenews.com/index.php?event=link,150&values[0]=1&values[1]=4063

7. http://www.globalresearch.ca/the-militarization-and-annexation-of-north-america

8. http://www.globalresearch.ca/it-s-official-the-crash-of-the-u-s-economy-has-begun, see also http://www.globalresearch.ca/financial-crisis-global-systemic-crisis-in-2007-another-bubble-close-to-bursting

9. http://news.bbc.co.uk/1/hi/6257194.stm

10. http://www.thesun.co.uk/sol/homepage/news/scottishnews/2046085/Doctor-Evil.html

11. http://www.alternet.org/story/62728/iraq_death_toll_rivals_rwanda_genocide%2C_cambodian_killing_fields

12. http://www.alternet.org/story/56124/is_the_united_states_killing_10%2C000_iraqis_every_month_or_is_it_more, see also post Gulf War I depleted uranium toxicity; http://www.ehjournal.net/content/4/1/17

13. http://www.reuters.com/article/2008/01/30/us-iraq-deaths-survey-idUSL3048857920080130

14. http://www.antiwar.com/ips/nazzal.php?articleid=12458

15. http://web.archive.org/web/20071107113350/http://www.thestar.co.uk/news?articleid=2984243

16. http://www.theguardian.com/uk/2007/jul/23/weather.immigrationpolicy1

17. http://www.telegraph.co.uk/earth/earthnews/3316004/Gordon-Brown-backs-nuclear-power-stations.html

18. http://baltimorechronicle.com/2007/112107Lindorff.shtml

19. http://www.globalresearch.ca/missing-nukes-treason-of-the-highest-order

20. http://www.ft.com/cms/s/0/ae209558-54a2-11dc-890c-0000779fd2ac.html, see also: http://www.voltairenet.org/article167754.html

21. http://gata.org/node/5738

22. IPCC; http://www.ipcc.ch/publications_and_data/publications_ipcc_fourth_assessment_report_synthesis_report.htm

23. CREDIT: IPCC; https://www.ipcc.ch/publications_and_data/ar4/wg1/en/spmsspm-projections-of.html

24. Source: Legget et al, *Global Warming: The Greenpeace Report*, Oxford University Press (1990), pg 20

25. Source: IPCC; https://www.ipcc.ch/publications_and_data/ar4/wg1/en/spmsspm-projections-of.html

26. Source: World Meterological Organization; http://www.wmo.int/pages/prog/arep/gaw/documents/ghg_bulletin_e.pdf

27. Source: J. Jaeger, *Developing policies for Responding to Climatic Change: a summary of the discussion and recommendations of the workshops held in Villach, 28 September to 2 October 1987*

28. CREDIT: UK Met Office; http://ukclimateprojections.metoffice.gov.uk/media/image/1/a/T_Fig1_large.jpg

Joy of Ambrosia:

1. Monbiot, *Heat*, Penguin (2006), pg xix

2. Such was my delusion of the time.

3. I am talking here about technical understanding of large scale problems and their possible solutions. Generally people in the developing world are more aware of local scale environmental problems than those in the developed world, as these environmental problems directly affect their day-to-day livelihood. Conversely, people in the urban developed world are largely insulated from the local scale changes people in the developing rural world are experiencing — but due to technology and literacy are more aware of the larger scale issues and possible technical responses.

4. Chussudovsky, *The Globalisation of Poverty and the New World Order*, Global Research Publishers (2003); see chapter 7 *Economic Genocide in Rwanda*

5. Lappe *et al*, '*World Hunger: Twelve Myths*', Grove Press (1998)

6. See Ponting, *A Green History of the World*, Sinclair Stevenson (1991), pg 107–9. Directing the relief effort on behalf of the British government was Charles Trevelyan — a student of Thomas Malthus. His sentiments on the Famine can be found quoted here:

"mechanism for reducing surplus population"

http://multitext.ucc.ie/d/Charles_Edward_Trevelyan

and here:

"The deep and inveterate root of social evil remains, and I hope I am not guilty of irreverence in thinking that, this being altogether beyond the power of man, the cure has been applied by the direct stroke of an all-wise Providence in a manner as unexpected and unthought as it is likely to be effectual. God grant that we may rightly perform our part, and not turn into a curse what was intended for a blessing." http://multitext.ucc.ie/d/Letter_of_Charles_Edward_Trevelyan_to_Thomas_Spring-Rice_Lord_Mounteagle and here:

"We must not complain of what we really want to obtain. If small farmers go, and their landlords are reduced to sell portions of their estates to persons who will invest capital we shall at last arrive at something like a satisfactory settlement of the country"

http://www.independent.co.uk/arts-entertainment/historical-notes-god-and-england-made-the-irish-famine-1188828.html

7. Lappe *et al*, (1998)

8. Chussudovsky, *The Globalisation of Poverty and the New World Order*, Global Research Publishers (2003); see chapter 6 *Somalia: The Real Causes of Famine*

9. See Altieri *et al* (1999) for results of organic sustainable agriculture in the developing world: *Report of Bellagio Conference on Sustainable Agriculture*, Cornell International Institute for Food, Agriculture and Development (1999)

10. Capra, *The Hidden Connections*, Harper Collins (2002), pg 76–8

11. Rose, *Dingo Makes Us Human: Life and Land in an Australian Aboriginal Culture*, Cambridge University Press (1992); see Chapter 13 *This Earth*

12. "Kapieren und Kopieren!" — "Comprehend and copy Nature!"; Viktor Schauberger's motto. For translations of his work see http://www.schauberger.co.uk/books.html

13. Lietaer *The Future of Money*, Random House (2001), pg 267

14. http://www.zeri.org/ZERI/Home.html

15. http://ourworld.unu.edu/en/biodiversity-in-the-hani-cultural-landscape

16. http://www.ecovillagefindhorn.com/

17. http://www.lietaer.com/2010/09/the-story-of-curitiba-in-brazil/

18. http://www.rmi.org/

19. http://www.lietaer.com/2009/12/
sustaining-cultural-vitality-in-a-globalizing-world-the-balinese-example/

20. https://notendur.hi.is/dagnyarn/Losun_grodurhusalofttegunda/2004-5_
Idntaeknistofnun_The_Icelandic_Hydrogen_Energy_Roadmap.pdf

21. Try the 'Contents of the Mind' Belief Exercise at the end of this book to begin exploring your own world.

22. See Wong, *Chi Kung for Health and Vitality*, Vermillion (2001)

23. CREDIT: http://earthobservatory.nasa.gov/NaturalHazards/view.
php?id=79800

24. Monbiot, *Heat*, Penguin (2006), pg 215

Bibliography

Corporate Globalisation

1. *Mortgaging the Earth* by Bruce Rich (Earthscan 1994)

2. *The Globalisation of Poverty and the New World Order* by Michel Chussudovsky (Global Research Publishers 2003)

3. *Confessions of an Economic Hit Man* by John Perkins (Plume 2005)

4. *The Best Democracy Money Can Buy* by Greg Palast (Plume 2004)

GAIA

5. *Revenge of GAIA* by James Lovelock (Basic Books 2006)

6. *The Web of Life* by Frijtof Capra (Harper Collins 1996)

7. *The Hidden Connections* by Frijtof Capra (Harper Collins 2002)

Money

8. *Healthy Money, Healthy Planet* by Deirdre Kent (Craig Potton Publishing 2005)

9. *The Future of Money* by Bernard Lietaer (Random House 2001)

10. *Islamic Banking and Finance in South-East Asia: Its Development and Future* by Angelo M. Venardos (World Scientific Publishing Company 2006)

Global Warming

11. *Global Warming: The Greenpeace Report* edited by Jeremy Leggett (Oxford University Press 1990)

12. *Global Warming: The Complete Briefing* by John Houghton (Cambridge University Press 1994)

13. *The Heat Is On* by Ross Gelbspan (Basic Books 1998)

14. *Boiling Point* by Ross Gelbspan (Basic Books 2005)

15. *The Rough Guide to Climate Change* by Robert Henson (Penguin, Rough Guides 2006)

16. *Heat* by George Monbiot (Penguin 2006)

17. IPCC website; http://www.ipcc.ch/

Terrorism

18. *Globalization and Terrorism: The Migration of Dreams and Nightmares* by Jamal R. Nassar (Rowman & Littlefield Publishers, Inc 2005)

Ecological Interdependence

19. *A Green History of the World* by Clive Ponting (Sinclair Stevenson Ltd., 1991)

20. *Collapse: How Societies Choose to Fail or Succeed* by Jared Diamond (Penguin 2005)

Peak Oil

21. *The Final Energy Crisis* edited by Andrew McKillop with Sheila Newman (Pluto Press 2005)

22. *The Last Oil Shock* by David Strahan (John Murray 2007)

23. *Half Gone* by Jeremy Leggett (Portobello Books 2005)

24. www.theoildrum.com

Political Power

25. *The Prince* by Niccolo Machiavelli (Penguin Classics 2003)

26. *Power* by Bertand Russell (Routledge 2004)

Geopolitics

27. *The Grand Chessboard: American Primacy And Its Geostrategic Imperatives* by Zbigniew Brzezinski (Basic Books 1997)

Philosophy

28. *A History of Western Philosophy* by Bertrand Russell (Unwin Hyman Ltd 1984)

29. *Indian Philosophy: An Introduction to Hindu and Buddhist Thought* by Richard King (Edinburgh University Press 1999)

30. *The Tao of Physics* by Frijtof Capra (Flamingo 1992)

31. *Sun Tzu: The Art of Warfare* introduced and translated by Roger Ames (Ballantine Books 1993)

32. *Dingo Makes Us Human: Life and Land in an Australian Aboriginal Culture* by Deborah Bird Rose (Cambridge University Press 2002)

Eastern Metaphysics

33. *Chakras and Kundalini: The Hidden Power in Humans* by Paramhans Swami Maheshwarananda (Ibera Verlag 2004)

34. *Chi Kung for Health and Vitality* by Wong Kiew Kit (Vermillion 2001)

35. *Autobiography of a Yogi* by Paramahansa Yogananda (Self-Realization Fellowship 1998)

36. *The Complete Book of Zen* by Wong Kiew Kit (Vermillion 2001)

Western Metaphysics

37. *Ask and It Is Given* by Esther Hicks (Hay House 2004)

38. *The Amazing Power of Emotions* by Esther Hicks (Hay House 2008)

39. *Seth Speaks* by Jane Roberts (Amber Allen Publishing 1972)

40. *The Nature of Personal Reality: A Seth Book* by Jane Roberts (Amber Allen Publishing 1976)

Changing Paradigms

41. *The Paradigm Conspiracy* by Denise Breton and Christopher Largent (Hazelden Information & Educational Services 1998)

Ethics

42. *Ancient Wisdom, Modern World: Ethics for the New Millenium* by His Holiness the Dalai Lama (Abacus 2000)

Belief Exercise: Contents of the Mind

Take some time alone and eighteen sheets of paper. Write the following headings on a blank sheet of paper; one sheet per heading:

The World, My Self, Energy, My Energy, Money, My Money, Power, My Power, Sex, My Sex, Health, My Health, Death, My Death, Peace, My Safety, Freedom, My Freedom

Then take one of these pages and — without editing yourself or worrying about any contradictions — fill the page with whatever statements come to mind on the topic of the page. Do this as quickly as you can without analysing what you have thought or written. Do this for all the topics, at separate sittings if necessary.

What you have just done is explored the world — not as it is, but as you believe it to be.

Don't like this picture?

Go to: www.agcusick.com for more information on how to bring about a conceptual revolution of your own foundation.

Resources:

Book:
www.taketheredpillandcureglobalwarming.com

Subjective Solutions:
www.agcusick.com

Objective Solutions:
www.ecologicalparadigm.com

Human Beings do not know who and what they are